KATRINA

MISSISSIPPI

Voices from Ground Zero

NancyKay Sullivan Wessman

For information contact
Triton Press/Nautilus Publishing, 426 South Lamar Blvd., Suite 16, Oxford, MS 38655.

ISBN: 978-1-936946-50-1

Triton Press
A division of The Nautilus Publishing Company
426 South Lamar Blvd., Suite 16
Oxford, Mississippi 38655
Tel: 662-513-0159
www.nautiluspublishing.com

First Edition

Front cover by Le'Herman Payton. Front cover photo from GettyImages.

Library of Congress Cataloging-in-Publication Data has been applied for.
Printed in the USA

10 9 8 7 6 5 4 3 2 1

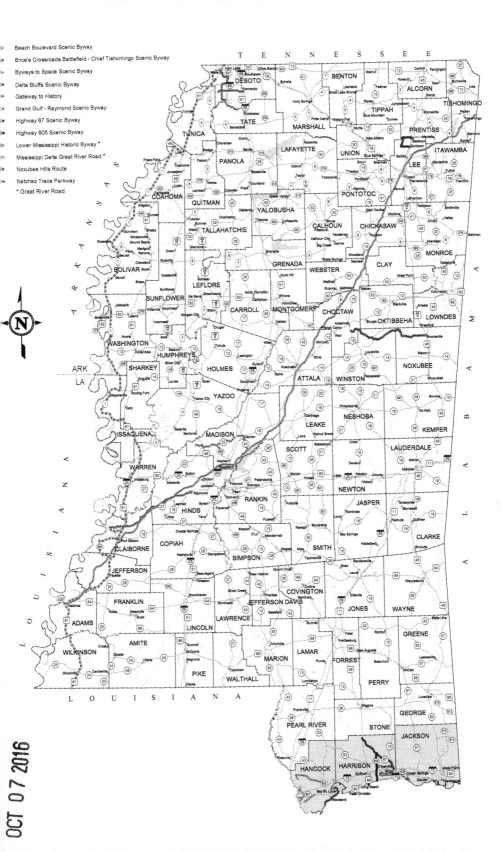

Beach Boulevard Scenic Byway

Brice's Crossroads Battlefield - Chief Tishomingo Scenic Byway

Byways to Space Scenic Byway

Delta Bluffs Scenic Byway

Gateway to History

Grand Gulf - Raymond Scenic Byway

Highway 67 Scenic Byway

Highway 605 Scenic Byway

Lower Mississippi Historic Byway *

Mississippi Delta Great River Road *

Noxubee Hills Route

Natchez Trace Parkway

* Great River Road

OCT 07 2016

Author's Introduction

As Hurricane Katrina developed off the west coast of Africa, built across the Atlantic, grew stronger over the hot Gulf of Mexico, and then hammered the Mississippi Gulf Coast, I watched and worried. For the first time in a quarter-century, I was not responsible for preparedness or response communications. I was retired from state service, no longer responsible for getting the word out before the storm and communicating important public health messages afterwards.

I had to remind myself, this hurricane was not my job.

Working as communications director for the Mississippi State Department of Health from 1979 to 2003, I was duty-bound to warn people about the danger hurricanes bring, the harm they can inflict, the debris and hazards they leave behind. What should individuals do to get ready? Should they move to a place beyond the storm's direct path? How could they deal with lost water pressure, thawed food, cuts and scrapes? Where could they get emergency medical care? My job required that I answer those questions, and on the response team I helped return the community to a better, safer, healthier state by ensuring, people had access to information and connections to disaster recovery assistance.

Before the turn of the century, before broadly available access to computers and the Internet, before smart phones and social media, we relied on communication tools and channels that today seem archaic. In the 1990s and early years of the 21st century, we relied on relationships with the mass media, on cooperation among all members of the public health team at every level, on the trust between government agencies and the citizens of Mississippi. Working with the Centers for Disease Control and Prevention and the National Public Health Information Coalition, for example, Mississippi's public health public relations team developed and hand-delivered to newspapers as well as radio and television stations a printed manual for disaster preparedness. Before hurricane season that year, we designed and gave the mass media a how-to guide for communicating from the public health perspective about hurricanes, ice storms, floods, infectious disease outbreaks. The guidelines, also given to strategic managers within the public health agency, contained "situation templates" for communicating objectives of that time: to strive for preparedness for bioterrorism, emerging infectious diseases, and other public health emergencies such as hurricanes, tornadoes, and winter storms.

Objectives reflected national and local concerns. Crisis response plans, procedures, and tools highlighted our changed worlds of practice after the Oklahoma City bombing of 1995 and the terrorist attacks of September 11, 2001. Those disasters cemented the critical importance of communication as an evidence-based public health intervention that must be considered before, during, and after an emergency. The responsibility to inform, educate, and empower people about public health issues holds a solid third place in American Public Health Association's list of "10 essential public health services."

After the terrorist attacks, the United States developed a National Preparedness System with a best practices-based National Response Plan for recognizing potential or real threats and for managing disasters. The National Interagency Incident Management System (NIIMS), an outgrowth of California wildfire-fighting tactics from the 1970s, would become the standard for all levels of government, nongovernmental organizations, and the private sector to "work seamlessly to prevent, protect against, respond to, recover from, and mitigate the effects of incidents, regardless of cause, size, location, or complexity." Particularly after 9/11, the federal government allocated hundreds of millions of dollars for responders' training and coordinated preparedness and response based on best practices. Mississippi embraced NIIMS as the State Incident Command System in 2001; the umbrella US Department of Homeland Security issued an enhanced National Incident Management System (NIMS) as the federal standard in March 2004; and in May 2005, Mississippi Governor Haley Barbour reaffirmed NIMS as the State standard for all emergency responses. *Before Katrina.*

Mississippi's public health system had embraced and followed that standard. Workers demonstrated that requisite coordination and cooperation among multiple local, state, and federal agencies through ice storms, the anthrax scares after 9/11, chemical spills, nuclear readiness, and the nation's then-largest pesticide contamination—methyl parathion, or cotton poison, misapplied in homes from Pascagoula, Mississippi, to Chicago, Illinois, in the 1990s.

But most of the people Katrina hit hardest neither knew nor cared about NIMS or any other government acronym. Then and now, individuals who most felt the storm's brunt simply strived to survive. Amidst the chaos of confidence against confusion, theirs became a hand-to-mouth existence.

Katrina grew, slammed into Mississippi's Gulf Coast, and demolished throughout the whole state. Katrina cost more than any other hurricane in US history and killed more people, directly and subsequently, than any since the 1928 Okeechobee hurricane.

New York Times bestselling author Douglas Brinkley devoted to Mississippi a chapter of his book—*The Great Deluge: Hurricane Katrina, New Orleans, and the Mississippi Gulf Coast*, released in early 2006. He called the storm a history-altering natural disaster. He wrote introductory notes: "My hope is that this history, fast out of the gates, may serve as an opening effort in Katrina scholarship, with hundreds of other popular books and scholarly articles following suit. . . Only by remembering, and holding city, state, and federal government officials responsible for their actions, can a true Gulf South rebuild commence in the appropriate fashion."

Brinkley's confidence challenged me, especially when the calendar changed to 2008 and I realized that nobody had told Mississippi's story, not really.

Biloxi *Sun Herald* published two books, *Katrina: Eight Hours That Changed the Mississippi Gulf Coast Forever* and *Katrina: Before and After*. Five years after the hurricane, University of Georgia Press published Pulitzer Prize-winning Poet Natasha Trethewey's *Beyond Katrina: A Meditation on the Mississippi Gulf Coast*. Also in 2010, Kathleen Koch delivered *Rising from Katrina: How My Mississippi Hometown Lost It All and Found What Mattered* (Blair), and Ellis Anderson offered *Under Surge, Under Siege: The Odyssey of Bay St. Louis and Katrina* (Mississippi). In 2012, University Press of Mississippi published James Patterson Smith's *Hurricane Katrina: The Mississippi Story.*

Post-Katrina reports came from various government entities, nonprofit organizations, advocates both local and from afar, and the mass media. But what *really* happened before, during, and after landfall to the people who owned the responsibility for managing the storm? Who were they, and how did they survive? The people who get paid to stay behind, the individuals who work to protect and serve through the local offices of government and related organizations—were they ready for what hurricane watchers predicted, "the next Camille?"

In his *Monster Storm*, Phillip Hearn reminded: "Hurricane Camille taught us lessons more than three decades ago, as did Hurricane Andrew just a little over one decade ago. The knowledge and experience gained from hurricane history, however, have not necessarily translated into changes in behavioral patterns or produced practical actions that would keep the disasters of the past from repeating themselves. Indeed, the nation's vulnerability to hurricanes increases in proportion to the increasing development of coastal areas. Short-term societal responses before and after hurricanes should be based upon decisions and policies made over the long term. As the Camille Project Report stated: 'Hurricane Camille, like every storm, provides a real-world test of the existing level of preparedness. Without exception, each storm reveals areas where society could have been better prepared or less vulnerable. If we are to identify those actions needed to improve a community's preparation for hurricane impacts, then we must focus attention on ways to ascertain a community's exposure BEFORE a hurricane strikes.'"

Hearn also wrote that, just a year before Katrina, no national hurricane policy existed, even though, "The stakes are certainly much higher now. Where and when will another monster storm strike? No one knows for sure."

Gulf Coast Mississippians now know.

The real mother of all monster storms struck August 29, 2005. And ten years later, the stories surrounding Katrina still bother many of the individuals whose jobs then centered on protecting and serving the 190,000 citizens of Harrison County and some 46,000 who called Hancock County home. Before Katrina hit, roughly 200 women and men hunkered in the Harrison County Emergency Operations Center (EOC) within the County Courthouse in Gulfport. Only 35 stayed in Hancock's EOC in Bay Saint Louis. They voluntarily stayed in that bunker and in that re-purposed bowling alley because they owned the responsibility for managing the unthinkable.

Katrina was not my job. My Katrina work began after the search and rescue, after the response, and well into the recovery and initial rebuilding phase. I first saw the infinite void she left—miles and miles of emptiness westward from Highway 49 in Gulfport to the Bay of Saint Louis and beyond—in January 2007. Most of the debris had been removed; nothing looked as it had before the storm. Even now, nothing looks the same: landmarks are gone, constructions that resemble bombed-out buildings in the world's war zones still stand in a few places, and new facades cover structures that miraculously kept their foundations and at least some of their bone structure.

Even the faces of the folks in charge then compared to those now responsible for emergency management have changed. Some simply moved on to other jobs or went back to their pre-storm responsibilities; several retired; others passed away.

But many who shouldered responsibilities within the EOC and outsiders who worked their way inside to help in the aftermath shared their stories with me. Through one-on-one interviews, opening their files, loaning their videos, referencing their materials—the characters had been waiting for the opportunity to tell their stories. They remained shell-shocked; some admitted post-traumatic shock; and they needed to talk, to find answers, to validate themselves.

Their stories stimulated me and compelled me to write.

Through the telling, I shall have turned from communicating the messages of preparedness *before* a natural disaster and of how to safely recover afterward to recording their portion of the history of the United States' costliest hurricane. This is one of the books Brinkley envisioned—a work of creative nonfiction that showcases heroes and their work from the epicenter of preparedness, response, rescue, recovery, and rebuilding. This account weaves their individual stories into a timeline that also reports events simultaneously occurring beyond the two Ground Zero counties—the accounts of state and federal governments' activities and the response of people and organizations from Florida to Oregon, Washington, and Alaska. This book deals with the public health impact of both the natural disaster and the unnatural consequences that emerged through human efforts. The book reveals personal recollections of health and medical aspects, special needs victims and mass care through sheltering, pop-up medical clinics, and the sole hospital that withstood the storm and continued providing services. The book introduces characters who addressed issues related to food and water, sewers, volunteers, donations, and other emergency support functions.

Readers learn of catastrophe and courage through the experiences of a public health physician, Robert Travnicek, MD, MPH, in upheaval not of his own making but caught in a quagmire of natural disaster, local and state politics, and moral determination. EOC Commander Benjamin J. (Joe) Spraggins, brigadier general, directs with able assistance from Rupert Lacy, a veteran law enforcement officer whose history, knowledge, and respect for the power of the storm enabled him to oversee logistics for all emergency support functions and, later, succeed his boss as emergency management director. Paramedic-elected-coroner Gary Hargrove set aside his own family's predicament to lead search and rescue, then recovery, and, finally, identification of each person Katrina killed in his county. And Steve Delahousey, veteran EMS leader on local and national levels, made sure special needs people were moved from harm's way before the storm and that adequate medical care was available after.

This book documents the players' personal and professional views as they reveal their alliances and actions, their concerns and issues, their truths and consequences. Theirs are stories about human suffering and survival—often in spite of assistance from government. The book does not distinguish right from wrong or comment on whether individuals or organizations succeeded or failed. Readers must draw their own conclusions.

These stories—the characters' perspective on the problems they encountered and what they themselves revealed to be their values through the storm of the centuries—can bridge to whatever becomes the United States' and the Gulf Coast's next Katrina.

Mississippi's Gulf Coast is rebuilding. The Coast will remain different from before the storm and distinct compared to most of the rest of Mississippi. Physical history, for the most part, washed away or was hauled into landfills. But the memories—as Mississippi's own writer of writers said—will persist because, especially in the Deep South, "The past is never dead. It's not even past." As William Faulkner affirmed, "I believe that man will not merely endure. He will prevail. He is immortal, not because he alone among creatures has an inexhaustible voice, but because he has a soul, a spirit capable of compassion and sacrifice and endurance."

- NKS Wessman

Prologue

The rest of the world had no way to know in the summer of 2005 that a hurricane of epic size and strength destroyed Mississippi. Getting to Gulfport and Biloxi requires extra steps, especially if the origin occurs outside the Magnolia State: fly into New Orleans or Mobile and drive to the Gulf Coast—a tri-county landmass that bargains every day with waters and wind from the South for its fundamental border. US Highway 90 and the railroad track parallel that mutable east-to-west line, and US 49 from its intersection at the foot of Gulfport all but centers the state northward through Hattiesburg, Jackson, and Grenada to Memphis.

That's the place—Harrison County, Mississippi—Dr. Robert Travnicek chose a decade-and-a-half earlier for his second career. He wanted the warmth, the charm of an historic Coastal town with a slower pace, and opportunities to make a difference from a public health perspective. And that's the place he struggled to save and rebuild through more than a hundred days after Hurricane Katrina exerted all the forces of nature against his chosen home base.

Robert G. Travnicek, MD, MPH, moved to the Mississippi Gulf Coast in 1990. He brought passion for public health, an outsider's perception of how to get things done differently, and 22 years' solid experience as a private family practice physician in Nebraska. A young University of Nebraska College of Medicine graduate, he also studied at Kansas and

Northwestern Universities as an undergraduate. The new medical doctor had envisioned a career in public health, but one month before he would have committed to the public health service branch of the US Coast Guard, his father died suddenly and prematurely. Even though Bob Travnicek had spent one seminal year in Mississippi as a federal assignee from the Office of the Assistant Surgeon General in the 1960s, he allowed professional and personal obligations to trump his personal desire. He went home to Nebraska, taking over the solo rural private medical practice.

"I've always been passionate about governmental medicine; public health was my opportunity to contribute," he said. "I was president of student AMA in medical school and have always been active in campus politics and organizational work. As a federal assignee from the Surgeon General's office, I was here the year after they dug those kids out of the dam in Neshoba County. I was here on Day One of Medicare, and I worked on a bunch of projects that were difficult to deal with. . . But I'm the son of a family physician in Nebraska who died suddenly at age 57; he was the only doctor in the area, and I felt I had to return to take up the slack there."

Two decades later, Travnicek once again left his small hometown community of Wilber, which proudly proclaims itself the Czech capital of Nebraska since 1963 and of the USA since 1987. He went to the Northeast, to Harvard School of Public Health, where, while earning the master's of public health degree, he also audited classes in the John F. Kennedy School of Government, the Business School, and Harvard Medical School. Practicing medicine was history; his future was public health.

He brought that new perspective to the place he chose, the Mississippi Gulf Coast. In that place, he became the first face of public health, the doc-in-charge for routine population-based health matters and the leading expert for all-hazards preparedness and response. And in that place, he learned from Wade Guice, director of Civil Defense in Harrison County for 35 years from 1961-1996, who spent most of his life preaching hurricane preparedness around the world. In that place in 2005, Travnicek firmly faced Mississippi's greatest threat, Hurricane Katrina.

Champions Of The Storm

- Brian "Hooty" Adam – director, Emergency Management Agency, Hancock County

- Henry Arledge – superintendent of education, Harrison County

- Linda Atterbury – director, emergency management, City of Biloxi, and deputy commander, Emergency Operations Center, Harrison County

- Haley Barbour – governor, State of Mississippi

- Raymond Basri, MD – volunteer responder, New York

- Michael S. Beeman – director, national preparedness, Region II, FEMA, and operations manager assigned to Harrison County

- Sherry Lea Bloodworth – volunteer responder, Fairhope, Alabama

- Jim Craig – director, health protection, and ESF-8 leader, Mississippi State Department of Health

- Michael Cruthird – director, early intervention program, District IX, Mississippi State Department of Health, and district liaison to ESF-8 Regional Command

- Steve Delahousey – vice president for emergency preparedness, American Medical Response, and ESF-8 co-director, Harrison County

- Greg Doyle – AMR EMT-Paramedic, and ESF-8 deputy director, Harrison County

- Jennifer Dumal – chief nurse and vice president for patient care, Memorial Hospital of Gulfport

- John Elfer – sergeant, Warren County Sheriff's Office, Vicksburg, captain, Mississippi National Guard, and military liaison officer to Harrison County

- Rick Fayard– AMR EMT-Paramedic, and ESF-8 director, Hancock County

- Myrtis Franke – staff assistant, Office of US Senator Trent Lott, Gulfport, and chairman, Board of Trustees, Memorial Hospital of Gulfport

- Diane Gallagher – vice president for marketing and planning, Memorial Hospital of Gulfport

- Eric Gentry – chief, field operations, FEMA, and operations manager assigned to Hancock County

- Wade Guice – director, 1961–1996 (*d, 1996*), Civil Defense, Harrison County

- Doug Handshoe – survivor, Gulfport

- Gary Hargrove – coroner, Harrison County

- Lawrence Henderson – vice president for administrative services, Memorial Hospital of Gulfport

- Tim Keller – chancery clerk and administrator, Hancock County

- Rupert Lacy – deputy sheriff, Harrison County, and logistics chief, Emergency Operations Center, Harrison County

- Robert Latham – director, Mississippi Emergency Management Agency

- Hal Leftwich – administrator, Hancock County Medical Center

- Gary Marchand – chief executive officer, Memorial Hospital of Gulfport

- Rick Martin – chief park ranger, Vicksburg National Military Park, lieutenant colonel, Mississippi National Guard, and military liaison officer to Harrison County

- Chip Patterson – commander, Northeast Florida Incident Management Team, Jacksonville, assigned to Harrison County

- Kamran Pahlavan – director, Utility Authority, Harrison County

- Brice Phillips – organizer and operator, WQRZ Radio, Hancock County

- Donald Rafferty – attorney, City of Bay Saint Louis

- Connie Rocko – member, Board of Supervisors, Harrison County

- Darren Scroggie, MD – volunteer responder, Mobile, Alabama

- Martin "Marty" Senterfitt – deputy commander, Northeast Florida Incident Management Team, Jacksonville, assigned to Harrison County

- George Schloegel – president and chief executive officer, Hancock Bank

- Joe Spraggins (Benjamin J. Spraggins) – director, Emergency Management Agency, Harrison County

- Tina Stewart – volunteer nurse, Emergency Operations Center, Harrison County

- Charles Stokes – president and chief executive officer, CDC Foundation, Atlanta, Georgia

- Deborah Taylor – nurse/infection control coordinator, Biloxi Regional Medical Center

- Gene Taylor – representative, US House of Representatives

- Robert Travnicek, MD, MPH – health officer, Coastal Plains Public Health District, Mississippi State Department of Health, and co-director, ESF-8, Harrison County

- Pam Ulrich – administrator, Harrison County, and finance chief, Emergency Operations Center, Harrison County

- Misty Velasquez – marketing director for tourism, Harrison County, and deputy coroner, Harrison County

- Bobby Weaver – director, Sand Beach Authority, Harrison County, and operations chief, Emergency Operations Center, Harrison County

Acronyms

ABC – Alcoholic Beverage Control
ADHD – Attention Deficit Hyperactivity Disorder
AKA – Also Known As
AMA – American Medical Association
AMR – American Medical Response
ATF – Alcohol Tobacco Firearms and Explosives, Bureau of
ATM – Automated Teller Machine
BFI – Browning-Ferris Industries (disbanded 1999)
CDC – Centers for Disease Control and Prevention
CDT – Central Daylight Time
CEO – Chief Executive Officer
CPA – Certified Public Accountant
DEQ – Department of Environmental Quality
DMAT – Disaster Medical Assistance Team
DMORT – Disaster Mortuary Operational Response Team
EMA – Emergency Management Agency
EMAC – Emergency Management Assistance Compact
EMEDS – Expeditionary Medical Support
EMS – Emergency Medical Services
EMT – Emergency Medical Technician
EMTALA – Emergency Medical Treatment and Active Labor Act
EMT-P – EMT-Paramedic
EMU – Emergency Medical Unit
EOC – Emergency Operations Center
EPA – Environmental Protection Agency
ESF – Emergency Support Function
FBI – Federal Bureau of Investigation
FAA – Federal Aviation Administration
FEMA – Federal Emergency Management Agency

HIPAA – Health Insurance Portability and Accountability Act

HURREVAC – Hurricane Evacuation (restricted-use computer program)

ICS – Incident Command System

IMT – Incident Management Team

IT – Information Technology

MACC – MultiAgency Coordination Center

MDEQ – Mississippi Department of Environmental Quality

MDOT – Mississippi Department of Transportation

MEMA – Mississippi Emergency Management Agency

MPH – Master's of Public Health

MRE – Meals Ready-to-Eat

MSDH – Mississippi State Department of Health

NASA – National Aeronautics and Space Administration

NASCAR – National Association for Stock Car Auto Racing

NHC – National Hurricane Center

NICU – Neonatal Intensive Care Unit

NIIMS – National Interagency Incident Management System

NIMS – National Incident Management System

NOAA – National Oceanic and Atmospheric Administration

NRP – National Response Plan

PEER – Performance Evaluation and Expenditure Review

POD – Points of Distribution

SLOSH – Sea, Lake, and Overland Surges from Hurricanes

SNF – Skilled Nursing Facility

USNS – United States Naval Ship

UCG – Unified Command Group

USAR – Urban Search and Rescue

WIC – Special Supplemental Nutrition Program for Women, Infants, and Children

Chapter 1

Joe Spraggins talked to God before he agreed to be the new civil defense director for Harrison County, Mississippi.

Hot sunshine and heavy humidity clung to the Gulf Coast throughout the summer of 2005. Benjamin J. (Joe) Spraggins, brigadier general, with thirty-plus years in the military and seven as base commander of the US Air Force/Air National Guard facility in Gulfport, decided to retire. That's when he asked God for a challenge.

"I didn't want to just be sitting still in my new career—I'd always had a challenge every day of my career, and I wanted to remain active. So I told God I didn't want to have a sit-down job and do nothing."

He turned down a defense contractor job that would have put him onto airplanes between Gulfport and Washington, DC. He said no to a similar federal liaison post with the Mississippi National Guard.

Late one Friday afternoon he answered the telephone. "Yeah, I'll take it," he told County Administrator Pam Ulrich. Then he turned and told his wife, "I've just accepted the job as director of Homeland Security and Emergency Management for Harrison County." Stunned, she reminded him that the county post was the least lucrative job he'd been offered. For seven months, he had shunned every opportunity. "Why?"

"I don't know why," he told her. Later, he reflected, "It just felt right. I knew long-time Civil Defense Director Wade Guice well, and I looked up to him—thought what a great leader he was. I had worked with him through the military through Hurricane Georges. I wanted to be like Wade. When they offered me the job, I told them my goal was to put a five-year plan together, to build what we would like for the EOC and operations. And I *won't* be around forever." When he met with Ulrich in early August to work out the details for the Board of Supervisors-appointed position, they picked his start date: Monday, August 29, 2005. "Needless to say, we didn't know what was going to happen that day. We had no clue!"

Six days before Spraggins was to begin that new career, local and national media reported a disorganized weather system in The Bahamas that could, but probably would not, pull its rainclouds into thunderstorms, and worse. That was, after all, the year that weather went wild. The season's storm stage was set, already hot from the globe's warmest year in more than a century and experiencing the 15th driest year on record in the United States. Environmentalists registered record rainfall in the Nevada desert, glaciers melting in Greenland, an unprecedented series of major hurricanes and tropical storms in the Atlantic, and the worst drought in a century in the Amazon rainforest.

Spraggins, retiring brigadier general, took note that the Atlantic disturbance "had some capability" and decided to report for duty as civil defense director five days early. Driving south toward the beach, he parked on the west side of the Courthouse and walked into the Emergency Operations Center (EOC) for the first time as commander.

By that Wednesday morning, county supervisors, city officials, and experienced hurricane responders had also begun to filter in, some with purpose and others just to see what the forecasters predicted. Sensitive to the season and the signs, locals expected trouble. Both the confidence based on a distinguished military career and a 52-year-old's energy for a whole new start allowed Spraggins to ease into his new role and listen to the surrounding voices of experience.

Access to that section of the Gulfport courthouse—just to the left from the west entrance—rarely requires more than pushing through the double doors. Spraggins' predecessor, Wade Guice, the county's civil defense director from 1961 through 1996, planned it that way. When the former courthouse burned in the 1970s and with the inestimable experience of Hurricane Camille behind him, Guice penciled off an 8,400-square-foot space that would allow emergency responders to enter an area designed specifically for the nature of their work: tight but adequate for 72 staffers within sight and hearing of each other while positioned to monitor local and national media via televisions overhead in the corner reserved for the county supervisors. Each responder would occupy an assigned seat at the tables arranged in a NASA-like operations room; each would have a telephone, other essential equipment, and supplies to simply sit and get to work. Guice intended that space to be a bunker, a defensive fortification designed to protect people and their work-tools from all hazards. Ancillary rooms provided for communications personnel, conferences, and private meetings for decision-makers. Beyond an inner hallway along the building's west wall were a small kitchen, bunks for both women and men, and restrooms.

For his part in response to the 1969 Category 5 Hurricane Camille, Guice earned international acclaim. That Hurricane-of-Record slammed into Mississippi's coastline with over 200-mile-per-hour winds and near— 35-foot-high tidal waves. She claimed 135 lives and left another 41 Coast residents forever missing.

Among other lessons, Camille taught residents of Harrison County that they needed a safe place to work during future emergencies. Like nine other counties in Mississippi, Harrison claims two county seats, one in Biloxi and another in Gulfport. Strategically located near the state's primary north-south corridor, US Highway 49, the Gulfport county seat contains the EOC. In 1977, politicians made sure in planning and construction that the First Judicial Courthouse could provide that safe space for the EOC. Intentionally, the blonde-brick courthouse claims concrete for frame, roof, and exterior walls that can withstand high winds. Clean, energy-efficient, and modern, the facility features highly polished Mexican tile flooring and open flow throughout its footprint. The first floor sits just over 26 feet above sea level and provides space not only for the EOC but also for offices of the Chancery Clerk, Tax Assessor and Collector, Purchasing, and Veterans Affairs. The Board of Supervisors conference center sits just inside and to the left of the east entrance. On the second floor are the courtrooms and clerks' offices, the Circuit Clerk, Law Library, District Attorney, and Sheriff's Office.

As Spraggins settled into the EOC August 24, 2005, he looked to Ivy Lacy as senior staffer. Since 1986, she had worked with Wade Guice and then his successor, Linda Rouse, who retired the previous April. As County

as his primary work place. He assured the new-to-the-job Spraggins and reminded supervisors and colleagues they had well-established procedures and protocols with well-trained personnel qualified to fill their various roles. "Harrison County has led the nation in preparedness since 1969. Some of us old dinosaurs are still floating around—we know what to do." All the locals could work together seamlessly to handle their responsibilities: transportation, communications, public works and engineering, fire and police protection, public health and medical services, and 10 other emergency support functions.

Thursday night, everybody in the Emergency Management Agency and at the EOC went home. Except Spraggins—he joined fly-buddies with the 187th Fighter Wing at the Air National Guard base for a long-planned Thursday-through-Sunday reunion.

About 7 pm, a weak Category 1 Hurricane Katrina landed along the east coast of Florida, killing nine and leaving massive destruction across the land. Then she veered south and west over the Everglades before entering the Gulf of Mexico. Downgraded to a tropical storm overnight, Katrina emerged from the peninsula to encounter the shallow, hot waters in the Gulf. Immediately, she strengthened and regained hurricane status, beginning to expand as she curved north.

Official meetings started Friday, with plans for the EOC to be fully staffed Saturday morning. Spraggins and his team watched the radar, got information from the National Hurricane Center (NHC), and fully expected some kind of storm event to affect the Gulf Coast within the next few days. At 1 pm Friday, the NHC predicted Katrina would continue northward and curve to hit Panama City.

To cover potential outer bands destruction along Mississippi's coast and beyond, Governor Haley Barbour declared a state of emergency. He activated the Mississippi National Guard and said the Mississippi Emergency Operations Center in Jackson would open the next day.

By Friday afternoon, what Spraggins called "the thing" came as projected into the Gulf. He and the Board of Supervisors along with other emergency management personnel decided they needed to make a move. That's when Delahousey and Travnicek took charge of medical and public health. Delahousey emphasized the point: "All disasters are local. According to the

County's emergency plan, Dr. Travnicek and I are in charge for medical and public health. Other disciplines come, and the federal government augments what we do. No matter how big the disaster is—even Camille—an important concept that even the feds will tell you is that all disasters are local. The feds, unless they're asked, will not come in and supplant—not unless they're asked."

Directly linked to the restricted-use, real-time data analysis computer program HURREVAC, which stands for HURRicane EVACuation, they agreed special needs people had to be moved. Spraggins and his team, already collaborating with Mississippi Emergency Management Agency staff in Jackson and with colleagues in adjoining Hancock County to the west, focused on safety for the special needs population. Together, they focused on safety for some 6,000 individuals in skilled nursing facilities (SNFs or nursing homes) or registered as home health patients, in addition to another possible 25,000 non-institutionalized persons with special needs. Travnicek and Delahousey communicated with the area's 11 hospitals and 15 SNFs, ordering the evacuation of four nursing homes.

Hospitals did not evacuate, but administrators released patients who could safely return home, clearing bed space for those who might need skilled nursing after the storm. Decision-makers worried about unnecessary re-location of hospital patients. "Just the sheer movement of people from intensive care units on ventilators—you'd think, 'Somebody's gonna die,'" Delahousey admitted. "That's why it's tough to make a decision to evacuate a hospital. If you've got 24 people on ventilators, somebody's going to die, even if the weather's good. Just getting them from one floor to another is a problem. Imagine having to carry the patients down stairs because the elevators don't work. You have to ask whether the benefits are going to outweigh the risk any time you're talking about health care facilities. Shelter in place is the key."

Warnings went out Friday afternoon. "Be prepared. Get your stuff; board up; check that generator out or get generators. Get supplies. Do what you need to do to be prepared because a storm can change any time."

Spraggins delivered that same "storm can change" message to his military friends at their Friday evening function. Everybody was happy, not really worried, waiting to see what would happen.

Within a few hours, Katrina's direction did change. By 10 pm, when the EOC was staffed about 40 percent, the Hurricane Center's prediction changed to show that the monster storm had veered further west and could dead-center Mississippi.

Lacy noticed about 11 pm Friday that the EOC was full of people who normally would not have been there. He attributed the human surge to technology—people watching satellite technology in their own homes and knowing that something big was about to happen. Representing law enforcement, the veteran volunteer and county employee went to the beach. "I went looking for The Weather Channel's Jim Cantore. The media hype is on—crashing white-capped waves make good television, and we know that Jim Cantore is a meteorological magnet. If it's a storm, and if it's a bad storm and Jim Cantore shows up, you pretty much know you're going to be at ground zero."

Knowing, too, that twenty-first century job and personnel requirements had far surpassed the operation center's capacity, staffers set up for press briefings in the courthouse lobby. The original conference room lacked sufficient space, but the wide hallways could accommodate chairs and enough room for the many TV crews to set up for doing live shots and multiple interviews at the same time.

At the same time EOC staff pushed boundaries outward beyond the operations center and into adjacent courthouse spaces to accommodate their needs, they limited access to the courthouse and, especially, the operations room, to non-essential personnel. They set up for sheltering-in-place, putting cots up in the hallways because they would be staying around the clock. They checked food supplies. They tried to get people to rest, but nervous anticipation affected everyone; nobody got more than catnaps.

They watched Katrina's path as they had her predecessor 36 years earlier, but no one believed any storm could ever be as bad as Camille.

In August 1969, as Hurricane Camille approached Mississippi, Civil Defense Director Guice begged people to evacuate vulnerable coastal areas. Although more than 130 people died in Mississippi, Guice got credit for saving hundreds of others. Just days after Camille, he addressed national media: "Tell people to leave a hurricane-prone area in plenty of time, and they will never become a personal victim of the tragedy you see here today."

Both Ivy and Rupert Lacy embraced Guice's get-out-of-here philosophy, but both had jobs that required them to stay and help protect the citizens of their home county. They were where they needed and wanted to be. Delahousey, Travnicek, County Coroner Gary Hargrove, and many other hurricane veterans waited with them.

Mayors and chiefs of fire and police from d'Iberville, Biloxi, Gulfport, Long Beach, and Pass Christian along with the Board of Supervisors also gathered at the EOC.

By 5 am Saturday, many had already set up at their duty stations—7:30 or 8 would have been the weekend norm. Nobody wanted to miss anything related to what they then saw as a Category 4 or 5 storm. Newscasters continued the "it's going to get worse" reporting, and the volume of incoming telephone calls exploded. Standing room only showed that local people saw the storm as a very serious prospect.

President George W. Bush declared a state of emergency in selected regions of Louisiana, Mississippi, and Alabama on Saturday, August 27, two days before Katrina's landfall.

"You'd never think that one community would be hit twice within 40 years with the same kind of catastrophic storm, but from the National Hurricane Center we know this is a bad one," Lacy observed to friends in the EOC. "She's showing all the characteristics, and we're going to have a bumpy next couple days."

Twice-a-day briefings began. News media, city and county officials, and other players were as ready as they would ever be.

Politicians met and decided to open shelters, even though their primary message via news media was "Go north. Get away from the water's edge. Get out of harm's way. Run from the water and hide from the wind. Go to high ground."

Press release number one on August 27 announced Sunday afternoon's opening of 17 shelters in the municipalities and Harrison County. "With the potential for tropical storm force winds beginning in our area tomorrow night around 8 pm, residents relocating to a shelter are encouraged to do so as soon as shelters open, during daylight hours. Residents are asked to bring these items to the shelter:

- Ready-to-eat food for three days

- Food and drink to sustain you and your family for at least the first 24 hours
- Medications and medical supplies
- Baby food, diapers, and other infant needs
- Pillows and blankets
- Battery-powered TV or radio
- Flashlight with extra batteries
- Books, magazines, board games, and toys."

The notice prohibited pets, alcohol, firearms, and smoking.

A separate release covered broader recommendations, strongly urging evacuations for all Harrison Countians. "If residents plan to evacuate north or northwest, they must do so immediately. The State of Louisiana will begin contra flow into Mississippi at 4 pm today.

- All boaters are encouraged to seek safe harbor while favorable conditions still exist. Boat owners should be aware that all local drawbridges will not operate to the open position once sustained winds reach 34 MPH.
- Patients with special medical needs who can leave the area without assistance should begin making preparations to evacuate immediately."

Saturday's third press release urged individuals with medical needs to evacuate "as soon as possible. Home confined patients with medical needs who can leave the area without assistance should monitor the storm closely and consider relocating away from the coast as early as tomorrow, depending upon the overnight progression of the storm's track. Look to friends and family members outside of the immediate coastal threat zones as a possible host shelter.

"Do not rely on public shelters or local hospitals as your point if you have medical needs. Plan to leave the area early and take enough medication and medical supplies to sustain your condition for at least 10 days.

"Kidney dialysis patients and other patients who require life-sustaining medical treatments should consult your doctor before making plans to leave the coast. Your doctor will help you find a safe and appropriate location away from the coast capable of caring for your needs."

Hurricane veterans inside the EOC knew that getting people to evacuate would be tough. Delahousey knew what had to happen. "Evacuation must be done 24 to 48 hours in advance, and that's a problem. The weather here today has been beautiful! It's difficult to convince 250,000 people to leave when the weather is pretty outside. But when the clouds start rolling in, and it becomes ominous, the people say, 'Oh, yeah—the weather is ominous. Yes, I'm ready.' But when the weather is good, no matter how much you're screaming on the radio or TV, they don't tend to take you seriously."

Four additional releases on Saturday announced the casinos' closings, availability of boarding for horses in Brandon in central Mississippi, the opening of two more shelters, and that Bayou Portage Bridge would be open every hour for marine traffic.

Katrina continued meandering through the Gulf, doubling in size and rapidly gaining grit and girth to taunt storm watchers as a Category 5 monster hurricane.

Chapter 2

Complacency, delay, denial, and pure dread—Rupert Lacy fought these demons throughout the weekend before Hurricane Katrina slammed ashore in south Mississippi.

Saturday arrived with bright hot sunshine, perfect beach weather, and a normal late summer's opportunity for play and community activities. Then clouds rolled in and rains poured. Intermittently, no rain, and more rain. Lacy described the day's offerings as "beyond typical weather." Even the seagulls had left, many reportedly having already winged it to Wiggins, nearly 35 miles inland. Before evening, Lacy heard himself urging Harrison Countians: "Go north, and leave early enough so you don't get caught in the hurricane syndrome of 2004 when people evacuated four times."

He remembered especially the turmoil Hurricane Ivan created: people trying to evacuate from that storm "sat in a traffic jam on Highway 49 going north. What would normally have been a three-hour trip to Jackson became a 12-hour wait. That created a problem for Katrina. People didn't want to get stuck."

More disturbing to him was that so many people who lived along the beach and on the backstreets seemed oblivious to the approaching storm.

"If you're going to stay, be prepared," he preached through media interviews to people along the Gulf Coast. "Have available three to five —emphasis on five—days worth of basic supplies because it's a bad storm. Have 30 days worth of medications; have cash . . . things we have learned over the years. We know that government will be here to help, but we just don't know when. And you can be sure that when Mother Nature wants to make a point, she's going to make a point."

Lacy spent most of the morning at the EOC.

At assigned seat 7 in the ESF-13 (public safety and security) position, Lacy represented law enforcement. Keeping an eye on overhead televisions and an ear for weather updates, he answered phones and tried to convince citizens to leave. "Kinda like the old timers," he told them, "I can go down and gauge what I see about a storm, whether I see anomalies. Where's the Sound right now? Where are the birds? What do we see? We're trying to get businesses to close. We know people are buying supplies, and some have boarded up. We want to open shelters only as a last resort. We're in a 'go' mode. We've got to be ready."

As the veteran hurricane watchers monitored the storm's overnight movement, County Coroner Gary Hargrove reported that his boat-racing friends in South Florida—the Sarasota area—simply lost a few shingles and a few trees when Katrina brushed over the peninsula. But even then, on Thursday evening, he had emailed his last to them: "I think we're going to get our ass handed to us." From then, he worked closely with General Spraggins to get ready, mounting preparations for the county to be able to rebound quickly. And he tried to prepare his family; he wanted them to leave.

"I lived nine blocks north of here when Camille hit. I did not want that to happen to me again," he confessed to fellow responders as they prepared. "My dad was a policeman who was called back to work, and I remember hearing the nails in the attic pulling our house apart. I've always said I would not stay for another big storm, but being in public service with public safety, then with the fire department as an EMT-Paramedic, and now as coroner—my commitment to the community has to be greater. I realize I cannot leave. And my family will not go."

Families across the Mississippi Gulf Coast and in other states also got caught up in the hurricane preparedness frenzy. Nearly 900 miles to the east and north, a federal employee visiting his son in Virginia that Saturday took note and prepared to join the Gulfport team for the most monstrous storm the US would ever have suffered. FEMA Region II National Preparedness Director Mike Beeman would meet Harrison County's EOC staff on Sunday.

Not officially in his new job until two days later, General Spraggins listened on Saturday, realizing he had not had the opportunity to train for what seemed all-too-familiar to some of his colleagues, the Incident Command System (ICS), a vital component of NIMS, the National Incident Management System. Spraggins paid particular attention to his team, who had trained and who knew from other storms what they needed to do two days out. They had already evacuated four especially vulnerable long-term care facilities and prepared the hospitals as best they could. They had ordered shelters to open and hotels and casinos to close by noon Sunday.

After, Spraggins was pleased that "we worked with them. The casinos had a shift change at 2 am; so we told them to start shutting down at 2. By noon, on the 28th, every casino and all the hotels on the beach had been cleared. They were super people—everybody bent over backwards to help us make sure all were ready. People were fantastic. From Saturday afternoon we did a mandatory evacuation of low-lying Area A, and then on Sunday morning, after we realized the storm was projected to come to Mississippi, we evacuated B Zone. We never mandated C be evacuated, and a lot of people have asked me why. We did recommend all leave and go to a safe shelter, but some of our hospitals are in Zone C; if we put a mandatory evacuation, we would have had to move the hospitals, and moving hospital patients would put people in greater harm than if they stayed."

Along with Spraggins, all the supervisors and mayors of the five municipalities collaborated on the evacuation and other decisions. A savvy soldier, Spraggins had begun building the new model of relationship and cohesiveness with each official when he applied for the emergency management job; long years in the military had taught him the benefit of teamwork, trust, and loyalty of the troops. By August 27, before he officially assumed command, he affirmed the group talked as "one entity. We are not talking as individuals because we want a 100 percent effort and unified front."

The Harrison County group conferred with Mississippi Emergency Management Agency (MEMA) Director Robert Latham in Jackson, heard from the National Weather Service in Slidell, Mobile, and Miami, and talked with Governor Haley Barbour and his staff—"all working as one to make sure we do not evacuate needlessly or in spots," Spraggins said. "We saw last year that crying wolf is not good. We had so many evacuations that people are having to make a choice whether to pay the rent or to evacuate. We've already had numerous opportunities to evacuate in 2005. We want to make sure we evacuate based on real need."

The team considered power, water, food supplies, transportation, and communications—essential functions the storm surely would impact. They ordered blue roof tarpaulins, ice, water, a Disaster Medical Assistance Team (DMAT), and port-a-potties to be stood up. Following the normal protocol for such requests to emanate from city to county to state to federal, Harrison County also called for coroners throughout the state to be available and for body bags to be ready to ship south. They alerted police and military assets because theirs would be taxed to the limit.

Satisfied to that extent, Lacy and Hargrove responded to a National Weather Service call from Slidell, Louisiana, that projected the Wolf River would crest at 23 feet.

"Our 500-year crest on that river, which happened in the 1990s, was 17 feet," Lacy told his friend. "We need to do a reality check."

Wolf River flows south from near Poplarville between the parallel Highways 53 and 603 and under I-10 to flow into Bay of Saint Louis, the estuary that separates Harrison and Hancock Counties along the Gulf Coast. Jourdan River on the bay's west and Wolf River along the east provide ample amenities for sunbathing on sandbars, kayaking and boating, fishing, walking trails, and both fish camps and home sites under historic oak trees and into the marshes.

Pointing along the river on a wall-mounted map, Lacy showed fellows in the EOC where he worked on marine patrol most weekends for civil defense, fire service, or the sheriff's office. He confirmed the projection via written message from Slidell and showed Spraggins. "This was the base elevation on that bridge," he motioned. "If we get 23 feet, on that bridge we would be standing in water up to about our knees. Look at that house; even in the attic of that two-story house, you're not going to survive. We've got to send messages out to the people using our Reverse-911 system. How far? I'd say from the mouth all the way up to Highway 53." Listeners noted the distance from Lacy's finger to the coastline, a good 12 miles by air and encompassing many more with all the river's twists and curves. "Let me do the message in my own words; I don't need a script. A lot of those people on the Wolf River know me because I'm on the river every Saturday and Sunday, and if they hear my voice I think they'll believe me."

Spraggins agreed.

Then Lacy recorded the message: "The National Weather Service is telling us we're going to have a 23-foot storm surge on the Wolf. Our historical river flood was 17 feet. So you don't have much time to lose. Get your property and go. Go now."

After sending that alert, Lacy and Hargrove drove to Henderson Point, the southernmost residential on-the-beach site fronting the Gulf of Mexico at the Bay of Saint Louis. With the sun setting, they encountered Deputy Herbert Roles, a former police officer and reserve member who "rode out Camille in '69 in the high school in Pass Christian, holding onto people and treading water." Greg Fredrico was there, too, but Lacy relieved him to go help with security at the Courthouse. Fredrico had already gone door-to-door insisting that all residents evacuate.

For those who refused to leave, Fredrico had already distributed and collected EOC forms that require name, address, phone number, emergency contact information, and a signature; he also gave out permanent markers and instructed each individual to write that same information on chest, *not* arm or leg, which could detach from a storm-destroyed body. He urged them to "leave. Get out now. We might not have assets to come down and rescue you."

Taking shelter from rain under the utility building at Henderson Point boat launch, Lacy said, "I don't think we're going to see this place like this ever again in our lifetimes. This is the highest I've seen this water in a long time—the side streets already have water, and people have secured their things only like they would for a typical storm and evacuated. But when we're able to come back, none of this will exist. I know my nickname is Doom-and-Gloom, but in emergency planning, you plan for the worst and hope for the best. This is going to be worse than what we can even anticipate."

With government assets stationed at three strategic sites within the county, EOC staff had begun pulling back from the waterfront essential tools they would need later: a boat and five-ton truck from Pass Christian, for example. Looking at the water marks already lapping the asphalt, they made the choice to save their equipment.

"The weather's starting to get bad," Lacy observed. "Looking out to the southwest from Bay Saint Louis, the clouds are just so not typically normal for a storm coming in. Those fast-moving clouds in the feeder bands are scary, and the storm's still a hundred miles off shore. The wind's picking up, the sea's swelling, the tide's rolling up onto the sand, and even the sea gulls are heading north."

Hargrove scanned the horizon and looked down at his boot, commenting that even the wharf bugs were lined up, headed north. He took some pictures. "I think this is the last time we'll be here and see this place like this. We're looking at 41 feet of water at the mouth of Wolf River; we're expecting a direct hit. I've got a bad feeling."

Lacy and Hargrove drove through Pass Christian, watching and marking the water. Heading east and north, back toward the EOC, Lacy noted some businesses had not closed. He muttered about their delay, their trying to make that last dime, and reiterated the urgency to leave. Some shops stayed open because locals still relying on Camille as the benchmark trickled in to get what was left of the bottled water, bread, peanut butter, toilet paper, tin fish, canned soup, vegetables, and fruit juices. They had already gotten flashlight batteries, coolers and bagged ice, and charcoal for grilling without power. The smart ones filled up with gasoline and got cash from the ATM. Some stopped by the liquor store.

Back at the EOC, Lacy tried to tell everybody what it looked like and that they all had to do whatever necessary to be ready. Delivering five life jackets to specific insiders, he all but begged police, fire, and other officials stationed to ride out the storm in Pass Christian, Long Beach, Gulfport, Biloxi, and even d'Iberville: "Take off all unnecessary equipment on your duty belts and keep only what you need—the gun, that's it. Secure the rest. Take off boots and put on lightweight gear unless you can swim in your boots. Take off your bulletproof vests. Wear reflective vests and put your whistle around your neck. It might look tacky, but it's a safety issue."

Lacy had already followed his own advice. Leaving his duty belt in the truck the day before, he worked with vest, gun, and whistle. Even now, he gratefully displays the whistle—"this one with 100 percent saltwater corrosion. Who would ever have thought?"

Back inside the bunker, Lacy reported in and caught up with what he might have missed. But he was restless. He knew some businesses continued to operate, even though casinos had begun their eight- to 10-hour process of shutting down. Built to withstand a 25-foot storm surge, the casinos were moving money, disconnecting services, and re-locating assets from the storm's reach.

"Something's not right. I can feel it in the pit of my stomach," he told his wife Ivy. "Something's eating at me. I'm running up to the house to check on some things one more time."

The gut feeling put him onto Highway 49 North about 11 pm Saturday. The drive to their home east of Saucier would normally take about 30 minutes. Even though he usually had seven sets of utility clothes in his vehicle for the unexpected, he did laundry—washing and drying everything not already cleaned, folded, and pressed. Then he telephoned Ivy. "What else might you need?" Before leaving, he put more dog food out. "Animals sense things. When I stepped outside, she was a magnet to me. She sensed something."

He thought about it on the way back to Gulfport. "I have never done anything quite like this, not since 1978. But tonight I have more than a gut feeling. Something's telling me. . . is it a sixth sense? Why am I driven to get everything done in this hour-and-a-half?"

After, he knew and could explain.

"I'm not the most religious individual in the world, but for several miles that night on 49 down to 53, I hit a tranquil moment. I had three types of police-public safety radios on in my vehicle, and the AM/FM radio's always on in my vehicle. I had the windows rolled down and the air conditioning on. But in that couple miles stretch, everything went quiet. Light rain had been misting all the way, but the rain stopped. There was no traffic except me. Nothing but my headlights in the quiet dark. If you ever have a point that you sense you have a guardian angel with you—it was like that guardian angel was there to physically and mentally prepare me for what was to come."

His wife heard his midnight-drive story first. Then he delivered the additional clothing into the EOC men's bunker before going back outside to sit on the curb. Telephone in hand, he began calling people he normally would talk to before a storm, probably a dozen friends. Many of them lived in nice homes at the water's edge.

"Leave," he urgently instructed. "Get into your car now. If you haven't done everything you need to do, get in your car now and go. Don't wait until in the morning to leave. Just get out of here now."

Chapter 3

While Mississippi responders monitored weather conditions that Saturday, the man who would become known as FEMA Mike also took action toward meeting the monster Katrina. Unexpected travel and talking to strangers come as naturally to Michael S. Beeman as a duck's waddling into water and swimming across a pond. Although he had not anticipated a quick trip to Mississippi in late August 2005, he took the assignment as only a veteran could.

Beeman directed national preparedness for Federal Emergency Management Agency (FEMA) Region II, which covers New Jersey, New York, Puerto Rico, and US Virgin Islands. When he got notification on Saturday, August 27, that he was needed in Mississippi, he was in Virginia visiting his son. In the face of Hurricane Katrina, FEMA headquarters wanted to know how quickly he could get to Jackson to handle Congressional public affairs and intergovernmental affairs.

Public affairs—he could handle that. From his first job as community relations advisor for the 57th Fighter Wing at Nellis Air Force Base, Nevada, Beeman had progressed through military newspaper editor posts at Rhein-Mann Air Base in Germany and at Kelly AFB in Texas through a 33-year string of increasingly demanding public affairs directorships from Florida to Sicily and back. As chief for external affairs in FEMA Region II, he had managed external affairs functions for 15 declared disaster or emergency response operations before moving up to the specialized niche in charge of preparedness issues as related to radiological emergencies, technical hazards, training and assessments/exercises. He worked to help state and local officials comply with the National Response Framework (NRF), focusing particularly to achieve compliance with NIMS.

The call to deploy that hot Saturday prompted his drive back to his home in New Jersey, where he got his Go Kit and drove to climb onto an airplane in Washington, DC, bound for Memphis, Tennessee. Having left leisure behind and landing in Memphis, he flipped on his cell phone and immediately recognized an urgent need to change plans.

"My Blackberry started going haywire with all sorts of notes," he would recall, most about the mayor in Gulfport wanting to have a FEMA representative on the ground before Hurricane Katrina's landfall.

As "most available candidate to go down there," Beeman flew into Montgomery, Alabama, where he'd asked staff to swap his pre-selected compact car for a more-suitable-for-hurricanes SUV. He intended to drive south toward Mobile and, finally, go westward to Gulfport. Having been through similar situations before, he stopped for gasoline and bought seven or eight cases of bottled water and other necessities before connecting with I-10 and heading into a situation he knew would be dire after the storm passed. Even though traffic was light—he'd encountered the east-bound stragglers evacuating from Louisiana and Mississippi's Gulf Coast about 10 miles west of Mobile—heavy rain from Katrina's outer feeder bands diminished visibility and slowed his normally hour-and-seven-minute trip of 75 miles to over two hours of white-knuckled driving.

Knowing that the Category 5 Katrina continued her laborious trek northward through the Gulf, Beeman showed no surprise when he exited onto Highway 49 South toward the Harrison County EOC in Gulfport. Except for one police car, nobody else was on the highway. People had already evacuated in the flood-prone areas and, if they stayed in the community, had sheltered in place. The SUV's navigation system took him directly to the courthouse and EOC.

Greeted at 8:30 pm Sunday by Gulfport Mayor Brent Warr, who'd requested pre-storm FEMA presence through the George W. Bush White House, Beeman met General Spraggins. Somebody sat in every assigned chair in the operations room—Rupert Lacy would recall that for the first time in all his years there, staffing was at 100 percent. The EOC was in near "lockdown," with every door into the courthouse except the west entrance secured. Hurricane shutters were down.

Representing city, county, and state agencies, assignees shared war stories and their own opinions about what they would face over the next 12, 24, and 72 hours. Some of the old-timers allowed that "we've been through Camille. We're in good shape."

"Well, yes, but," Lacy cautioned, "the Coast wasn't built up in 1969 as it is now." Construction, commerce, casinos, and even the culture differed dramatically from the pre-Camille coast.

Beeman turned the conversation back to what the responders already had accomplished. They listed supplies they had ordered, and he ordered more. He met separately with political figures and then briefed the team. Confident the locals had satisfied the National Incident Command System (ICS) checklist, Beeman began to relax.

When the Gulfport mayor mentioned that Weather Channel Meteorologist Jim Cantore wanted to interview him, Beeman said, "Well, take me down there to where he is." Driving along scenic Highway 90 for his first time and toward the Armed Forces Retirement Home, which he would remember as the "old soldiers' home," Beeman did not see Cantore. But he did stop to admire the beauty of lushly landscaped homes surrounded by centuries-old live oak trees—the sparkling white wood-frame or brick homes there since the early twentieth century, many having survived the historic Hurricane Camille.

"I knew I had to look, observe, and remember," Beeman recalled later. "I said, Mr. Mayor, take one last look at it. This will not be this way tomorrow. All this will be destroyed."

"You're kidding!" the mayor said.

"No, no," Beeman quickly responded. "You're looking at a Category 5 storm. All this will be destroyed and gone tomorrow."

That story later became a primary start-point for the mayor's telling of the storm, and Beeman himself heard Warr telling the President at least once, if not twice, "'That FEMA guy you sent down here took me down and told me what was going to happen. Darned if he wasn't correct about everything that happened.'"

Without locating Cantore then, Beeman separated from the mayor and went to his overnight accommodations at the Air National Guard base at

KATRINA IS MOVING TOWARD THE WEST-NORTHWEST NEAR 10 MPH. A
GRADUAL TURN TOWARD THE NORTHWEST IS EXPECTED LATER TODAY.

MAXIMUM SUSTAINED WINDS ARE NEAR 145 MPH WITH HIGHER GUSTS.
KATRINA IS A CATEGORY FOUR HURRICANE ON THE SAFFIR-SIMPSON
SCALE. SOME STRENGTHENING IS FORECAST DURING THE NEXT 24
HOURS.

HURRICANE FORCE WINDS EXTEND OUTWARD UP TO 85 MILES FROM
THE CENTER...AND TROPICAL STORM FORCE WINDS EXTEND OUTWARD
UP TO 185 MILES.

ESTIMATED MINIMUM CENTRAL PRESSURE IS 935 MB...27.61 INCHES.

COASTAL STORM SURGE FLOODING OF 15 TO 20 FEET ABOVE NORMAL
TIDE LEVELS...LOCALLY AS HIGH AS 25 FEET ALONG WITH LARGE
AND DANGEROUS BATTERING WAVES...CAN BE EXPECTED NEAR AND TO
THE EAST OF WHERE THE CENTER MAKES LANDFALL.

RAINFALL TOTALS OF 5 TO 10 INCHES...WITH ISOLATED MAXIMUM
AMOUNTS OF 15 INCHES...ARE POSSIBLE ALONG THE PATH OF KATRINA
ACROSS THE GULF COAST AND THE SOUTHEASTERN UNITED STATES.
THE HURRICANE IS STILL EXPECTED TO PRODUCE ADDITIONAL
RAINFALL AMOUNTS OF 2 TO 4 INCHES OVER EXTREME WESTERN
CUBA...AND 1 TO 3 INCHES OF RAINFALL IS EXPECTED OVER THE
YUCATAN PENINSULA.

ISOLATED TORNADOES WILL BE POSSIBLE BEGINNING SUNDAY EVENING
OVER SOUTHERN PORTIONS OF LOUISIANA...MISSISSIPPI...AND
ALABAMA...AND OVER THE FLORIDA PANHANDLE.*

Emergency preparedness officials at MEMA headquarters in Jackson and from neighboring Hancock and Jackson Counties conferred via telephone with Harrison County EOC. All heard the latest from National Weather Service, the National Hurricane Center, and the governor.

"Guys, y'all make the decisions, and we'll back you in whatever you need," Governor Haley Barbour assured them.

After the conference call, Spraggins said to his preparedness staff, "We know we're going to get hit. And it's going to be one huge storm. The eye has started spreading; the storm's started moving. We've got to beg people to get out."

Video conferences with President George W. Bush and National Hurricane

Center Director Max Mayfield underscored what Mayfield called a "very, very great concern." He indicated the sheer size of Katrina and extended warnings to cover most of the Louisiana coastline and a larger part of the Florida Panhandle along with the entire coasts of Mississippi and Alabama. He predicted the area ultimately hit would be devastatingly damaged and "uninhabitable for weeks." President Bush announced Sunday that he had issued disaster declarations on Saturday for Mississippi and Louisiana. Bush begged people in the hurricane's path "to put their own safety and the safety of their families first by moving to safe ground."

With the help of radio and television, Spraggins reiterated the evacuation orders: "Pack your stuff; get out of here; go now. Get to safe haven, knowing that safe haven doesn't mean you've got to get in your car and go 200 miles north. Get out of the direct impact of the wind and direct impact of the flood, the water, the surge. If your home is on the Harrison–Stone County line, and if it's not a mobile unit, your home is probably about as safe as anywhere else. Stay there; buckle up; tighten it down. Know that we can't predict tornadoes—they're going to spawn off, and we cannot predict them. Be vigilant. But if you're in Evacuation Zones A or B, remember that just because your home survived Camille, don't expect to be okay in this storm."

He tried to explain the danger of the anticipated storm surge.

"When you're within a hundred miles of the coast and the winds are 160 miles per hour, you've got a 160-mile-per-hour surge. The surge is coming; please be prepared. The Coast is not the same as it was in 1969. We have more infrastructure, and it's going to cause more damage. The surge will be bigger; it's a bigger storm. The Coast is not the same; expect more flooding because during 35 years of progress, you've had to build a lot of things. Roads, drainage, everything has changed. If you take any area in downtown Biloxi, Gulfport, or Long Beach—those areas have changed. Some of the drains have been covered up or clogged up, and the water has no way to get out and will keep building and building and building. That's what we've got to tell people: if you lived through Camille, you might not live through Katrina."

Rupert Lacy underscored the weight of Spraggins' warning: "It's almost too late. We know asphalt does not absorb water, and concrete does not absorb water. The more development we've had, the more growth we've had, it affects us; it hurts us. If you don't have a barrier to stop water, it's going to take that path of least resistance, and that's concrete and asphalt. If you've decided to ride out this storm, think again. Leave."

Spraggins sent fire- and policemen to talk people out of their homes, knowing that mandatory evacuation cannot force anyone from his or her home; only businesses must obey that order. At 1 pm Sunday, he approved news release number five that urged, but did not order, evacuation for Zone C.

The Harrison County Board of Supervisors **strongly urges** *all residents of Harrison County to evacuate, including Zone C. The Board also urges residents countywide who must stay to use shelters north of I-10.*

The only reason *the evacuation of Zone C is not mandatory is to protect patients who are in hospitals in that zone. The Board has decided there is greater risk to move the patients than there is to keep them in place in the hospitals and the Board will not put the patients in harm's way. If the patients were not there, evacuation from Zone C would be mandatory.*

The Board urges use of shelters north of I-10. *Shelters south of I-10 should be used only as a last resort, and the Board discourages using them.*

A **curfew** *goes into effect within the Biloxi city limits at 7 pm and for the rest of the county at 9 pm today (Sunday, 8/28/2005). Curfews will remain in effect throughout the storm until damage assessments have been made and officials declare the roads safe for travel.*

Mandatory *evacuation orders remain in place for Zones A and B. Mandatory evacuation also applies countywide to mobile homes, buildings that cannot withstand strong winds, and harbors.*

A mandatory evacuation is a legal order to all people in affected areas, directing them to leave those locales and seek shelter elsewhere.

All shelters in Harrison County opened at 12 pm today.

The Board warns citizens: *Law enforcement, fire, and paramedic service will not be available during high winds. High water may prevent rescuers* *from reaching citizens in need. Further, all drawbridges will not open once sustained winds reach 34 mph.*

Hourly meetings at the EOC continued to focus on essential needs for both during and after the storm: food, water, supplies, fuel. Rupert Lacy remembers that Sunday as "the god-awfullest. Everybody was wired. We were 100 percent staffed; somebody was sitting in every chair—some were there for back-up, but we were full. We were in lockdown. Everything here was locked down except the west door; we had hurricane shutters down on most everything. As we got information from National Hurricane Center, we were digesting and getting the information out to people."

Based on Wade Guice's plan and the routine for every disaster's activation in Harrison County since 1977, the Board of Supervisors and/or city officials occupied assigned seats curved around the far corner, sitting under overhead television screens. Seats one through six offered work space for Connie Rocko, Larry Benefield, Marlin Ladner, and William Martin as well as for Richard Rose of d'Iberville and Rick Weaver representing the City of Gulfport. Seats seven through 16 were assigned to and occupied by law enforcement and hazardous materials: notably, Sheriff's Deputies Rupert Lacy and Greg Fredrico; fire services representative Bruce Wilkerson, who also acted for the City of Pass Christian; the gaming industry's state regulators; the Office of the Attorney General; public works and engineering; and human services, mass care, donations/volunteers, and shelters, under management of the Red Cross. Cramped into the curved third tier were medical and public health assignees and the military: EMS, Department of Health, Navy, Air Force, and National Guard. Public information officers claimed seats 27, 28, and 29—assigned to but not filled by energy, animal control, and Corps of Engineers—in the back corner. Spraggins directed from his office assigned position 30, and MEMA's Tom Taylor worked from Seat 31 among the four chairs at the front, in the corner, near the supervisors.

From the medical and public health arena, Steve Delahousey felt both calm and concern. Looking at all the hospitals and nursing homes located north of I-10, he was satisfied their evacuation of near-to-shoreline facilities had been the right call. Manning his station in the EOC and comprehending the situation as he did on Sunday stimulated serious concern: "Oh, shit! It's going to be bad!" he predicted to others also working the telephones. "The Highway Patrol and the National Guard are going north on Highway 49. They're waving goodbye to us. We should leave; we're in the flood zone—but we can't! Everything pre-deployed down here is headed for Camp Shelby." In the aftermath, he grimaced and shook his head: "It was absolutely the right thing to do."

Delahousey's next-seat-neighbor, Dr. Robert Travnicek, returned to his desk a little after 4 pm Sunday. He'd been in and out many times since storm-watchers began their vigil the previous Wednesday. As district health officer with Mississippi State Department of Health and responsible for managing the six coastal counties—Hancock, Harrison, Jackson, Pearl River, Stone, and George—and headquartered in Gulfport, Travnicek spent most of those first three days securing public health resources. He instructed district-level supervisory staff over personal and environmental health services as well as those responsible for finances and facilities, reminding them to review their agency's emergency response plans and to get ready; he told them what to expect and implored individuals to take all precautions for themselves and their families. On Saturday, he battened down his own property north of the railroad tracks in Long Beach. Driving his wife Lora, a nurse at Garden Park Hospital, to her workplace for lock-down mid-afternoon Sunday, he talked about the storm's force that would destroy everything in her path.

"This storm is potentially the biggest thing that's ever hit this country," he predicted, assuring his wife. "Wherever she hits, public health and medical systems will be particularly stretched. But I think we're ready—as ready as anybody can be. We've had at least one, or the threat of one, storm a year for the 15 years we've been here. Everything which is about to unfold and the potential consequences we must deal with in only medical terms—it's going to be the highest form of the practice of medicine. Good thing I've had meticulous preparation at some of the largest medical institutions in this country."

With spouse safely delivered and his vehicle stashed on a higher level of the Hancock Bank parking garage several blocks from the EOC, Travnicek settled in for what he knew would be a long, tiresome night. Telephones rang incessantly, and as night fell, the center became more crowded. He gazed around. "All newcomers," he thought from the physician's perspective. "Delahousey and I are the only two people who actually know what's going on here—almost everybody else is new, and few have any experience in the EOC except him and me."

Delahousey all but grew up in the EOC. A Gulf Coast native, he had spent his entire professional career in the health care industry, specifically emergency medical service (EMS). A registered nurse and registered emergency medical technician–paramedic, he had served four consecutive governor-appointed terms on the Mississippi EMS Advisory Council, chaired the state's medical control committee, and also served as chairman of the Harrison County 911 Commission.

Both men recognized the import of their roles as designated medical and public health team leaders. "State law grants pretty extraordinary powers under times of emergency," Delahousey said. "Sometimes you have to make decisions, and you don't have time to collaborate, make conference calls, and gain consensus; sometimes you've just got to act."

During those final build-up hours, General Spraggins focused his team on who they were, what they were to do, what they aimed to achieve. Every representative had a reason and a responsibility: port-a-potties, generators, tents, blue roofs, hospital patients, water, transportation, food; each had a different objective. At Spraggins' prompting, each verbalized his job and desire.

"You need to know who's beside you," the commander said. "We need to know who's here and how each of us will operate, to get acquainted with jobs and expectations. We need to know who the go-to people are. . . We can eliminate time wastes on the other side of this storm."

Ever observant with mind targeted on strategy, Travnicek spoke: "I want to work with you to avoid social unrest." Individuals stared at their neighbors, questions obvious in their eyes. Even Spraggins seemed perplexed.

"Here's the thing: public health cannot function in social unrest," Travnicek explained. "I can tell you from experience—for example, from my time at Los Angeles County General Hospital during the Watts Riots in 1965—that people milling about, agitated, uncertain, afraid, physically restless, and even disoriented can turn into a crowd feeding on rumor and uncontrolled behavior. Disruption of the norm could result in frenzy, looting, and possible violence."

Thoughtful silence ensued. But the telephones kept ringing.

County marketing director for tourism Misty Velasquez had reported for work at 6 pm. Having handled shifts for the previous two days, she worked the phones. "People are stuck in their homes; they are trying to find their families; or they are stuck outside the area and want to know what's happening here."

By the time Mike Beeman with FEMA arrived later Sunday evening, the emergency responders had already sent supplies requests up the chain to MEMA. Lacy had polled the cities to learn their needs and added those to the county's petitions. For the first time, he saw a federal agent in action. "As requests went in, he could increase the quantities. He asked us how many portalets we'd requested, and he added 400; finally the order got up to about 1,500. He didn't really let us know the FEMA assets that were being stood up. Their plate was very full. They had people on the ground from Alabama to Texas; they, too, had to report in—there was a lot of behind-the-scenes activity going on that you could never imagine. And people still don't know about it. Unless you're in the saddle dealing with it, you don't know what's going on. . . . I never had paid that much attention to the federal side."

Forecasters said Katrina would be a daytime storm, but Spraggins and his team were almost too nervous to wait. Some of them did not wait.

Near midnight, hours before the expected surge, Lacy, Fredrico, Spraggins, and Supervisor Rocko went out to look. "We went out to watch the water," the EMA director would later explain. "God love him, Rupert knew where every water line was for every storm since back to 1925. He knew. So we rode around that night learning from Rupert where previous storms' water had gotten. He knew it like the back of his hand."

Spraggins brushed away the danger. "It really wasn't that bad. Probably around midnight, we were on the beach, and water was coming up to the edge of Highway 90, overtaking vehicles even then." They saw no other people but did notice a vehicle here or there—one huge new pickup truck they had observed the previous night was already flooded, washing itself in and out of the Gulf. "The water kept coming, the water kept coming, the water kept coming, and the water kept coming. And the wind got up a little more."

Lacy had aimed his assigned vehicle, a fancy new Tahoe, toward the beach, intending a westward trek to Henderson Point. But not far from the intersection of Highways 49 and 90, he could see that water already swept over Broad Avenue. He pulled back, telling his passengers that what they saw was "impassable. But that's not the surge. It's just mean sea level water the Gulf has churned up. Still three or four hundred miles out, she's going to churn those seas up and water has to expand to go down; so it's pushing it out. All that water that makes this a Category 5 storm has to go somewhere; and it pushes until it gets in contact with land."

Lacy parked momentarily—just long enough to measure wind speed near the Port of Gulfport at the intersection of Highways 49 and 90. Then he drove east along Railroad Street, knowing that from Courthouse Road he could see from another angle. As they crossed Hewes Avenue, Lacy indicated the marine office for Sheriff's Office supplies. "We've got a 14-foot Boston Whaler in there, decked out. We're going to pick that boat up on the way back and park it inside the courthouse, in the lobby."

He stopped on a bridge near the VA Center, looked to see water lapping at the base, and declared "a big problem. Somebody could drown in this steady stream, and it's getting higher. What this tells me right here is that down there on Highway 90 in front of the VA Center, even on the northbound lane, you've got a foot-and-a-half of water taking that lane over; so that road is impassable, and it's got a couple dips through there that could drown somebody. If you're right by the creek—that's a big box culvert and with that water—that could push a car."

They drove past the old courthouse and the VFW, to where they could see the pier area. "Look at that truck parked over there. What kind of fool?" Two policemen had already confirmed nobody was inside, supposing somebody probably had dumped it there for what they called an "insurance case."

Lacy called watching the water from there "odd. It's coming in from the east quicker and rising more than from the west." He left the SUV's engine running, foot on brake, hands on the steering wheel, prepared to throw it into reverse. "White cap after white cap! Some of those waves are four or five feet, and with the wind gusts, sometimes six. We're talking for every 10 miles an hour a one-foot wave. Have y'all seen enough?"

For the trip that probably took about 30 minutes but seemed like hours, the three could look only at the water. Near the marine supplies building, they agreed. "Screw the boat. Let's get back."

Hargrove also went out about 4 am Monday, taking with him the general, a representative from the health department, and Misty Velasquez, a fellow county employee and friend who shared his enthusiasm for boat racing. Driving west to Highway 49 and then south to 90, he pointed, "Wow! Look at that—here's a 41-foot sailboat right there at the intersection on Highway 90."

"What about that one sitting *on* Highway 90?" Spraggins asked.

Hargrove looked. "Oh, my goodness." Blocked from driving east, he turned west. At 30th and 90, he stopped to take a picture of the Grand Casino parking lot. The water was already lapping the northernmost lane of Highway 90. He drove toward the traffic light at the entrance to the casino, took another picture, and drove another short distance before putting the window down and shooting a final photograph. Without really looking at the scene, he saw the picture and felt scared. "That little park they built out there under the oak trees is totally under water. The water's coming up all the way to our roadway. We need to go, and we need to go now. Seriously, we need to leave now." He drove away as hurricane winds blustered against the nearby hotel. Water rose quickly.

Back at the EOC, Hargrove continued his relentless checking for new information, talked to people nearby, and telephoned to check on his parents. Already having asked for DMORT (Disaster Mortuary Operational Response Teams) to be put on standby, he told Spraggins and the medical examiner's office that he probably would need more help than he had anticipated. Several years earlier, he had institutionalized an emergency preparedness form required of anyone who refused to evacuate when ordered. In filling out the form, the individual relinquished the county of responsibility if anything tragic happened and also provided personal contact information for family or funeral home in case they died. In media interviews about evacuation orders, he minced no word. "Everybody needs to heed the warning. I was here during Camille; so I know what happened, and I don't want that to happen to us again."

Tensions inside increased along with the wind and the rain outside. "We're going to get hit hard," Hargrove told them. Then he went with Lacy and one of the guardsmen into the Supervisors' Board Room, where he climbed onto a chair to peer out the window, placed high on the wall. "We're in trouble—look at the water on the road. It's flowing north, not south. We need to go see the general." Leaving that room, they noticed water on the floor of the Courthouse lobby. Behind closed doors, they reiterated their fear that water would come over the railroad tracks, which would require evacuation of EOC personnel to the second floor. That became the contingency plan.

The slow-moving hulk of a storm emitted ever stronger winds and dumped more rain. Katrina was closing in. About 7 am Monday, Deputy Lacy locked down the EOC as a hurricane shelter, finally closing the west entrance, too.

"Then we lost power," Hargrove recalled. "The steel hurricane-proof doors were down, and we couldn't get out. Then the generator went down, and we had to get somebody outside to get it going. We did get a door open, and, of course, the calls were coming in steady: People needing help, people taking on water in their homes, people telling us their house was totally under water, giving us addresses."

With the steel storm shutters down and secure, Lacy knew one way out: the north fire exit, partially protected from the forceful winds but with water already in the stairwell. Moving cautiously through the totally dark complex—with candles and only a couple flashlights, he made his way outside the building with an emergency management operations technician, a Department of Corrections inmate assigned to him, and a National Guardsman. "We're talking water—we were an island," he remembered. "They jerry-rigged the generator while I took another couple inmates assigned to our building and grounds crew. We started opening secure areas upstairs in case we had to evacuate from the EOC—just three feet of water would affect our computers. It was scary, and it seemed like an eternity, but it was probably only 15 to 20 minutes."

As the daytime storm banged the bunker, EOC principals tried to focus on the future to mitigate their own fears of the present; they discussed when they might safely emerge from the bunker to search for and rescue the callers. Sixty-five mile-an-hour winds would be safe, they decided. During the interminable wait, Hargrove called home again. "What's that?" he asked his wife about the background noise. "She said to hold on, stepped out to the patio, and then came back and said the end of the house was coming off. I told her to get to the center of the house, close to something that will protect y'all. Then she said the other end was coming off, and I said, 'Y'all need to get to the center of the house and stay put—away from the windows!' I got that out, and the phone went dead. My first instinct was to go jump in the car, but I was quickly told, 'No. You know you can't get there. We're surrounded by water.' . . . It was Thursday evening before I really knew they were okay."

National Weather Service would report that the Category 5 monster packed sustained winds around 175 MPH on August 28 and trekked north to land along the Gulf Coast early Monday as a large Cat 3.

National Oceanic and Atmospheric Administration's National Climactic Data Center recorded Katrina's second and third punches to the United States and followed her devastation of Mississippi:

> Though winds near the center had dropped to 150 mph, gusts to hurricane force were occurring along the coast. NOAA Buoy 42040, located about 50 miles east of the mouth of the Mississippi River, reported a peak significant wave height of 55 feet at 06:00 CDT, which equals the highest ever measured by a National Data Buoy Center (NDBC) buoy.
>
> At 06:10 CDT, Katrina made landfall in Plaquemine Parish just south of Buras (between Grand Isle and the mouth of the Mississippi River) as a strong Category 3 storm, despite entrainment of dryer air and an opening of the eyewall to the south and southwest. Landfalling wind speeds were approximately 127 mph with a central pressure of 920 millibars—the 3rd lowest pressure on record for a landfalling storm in the U.S. Winds at this time were gusting to 96 mph at the Naval Air Station at Belle Chasse, LA and to 85 mph at New Orleans Lakefront.
>
> By 08:00 CDT, Katrina was only 40 miles southeast of New Orleans with hurricane force winds extending outward up to 125 miles. In the dangerous right front quadrant of the storm, Pascagoula Mississippi Civil Defense reported a wind gust to 119 mph and Gulfport Emergency Operations Center reported sustained winds of 94 mph with a gust to 100 mph. New Orleans Lakefront reported sustained winds of 69 mph with gusts to 86 mph. A little

earlier, Belle Chasse reported a gust to 105 mph.

By 10:00 CDT, the eye of Katrina was making its second northern Gulf coast landfall near the Louisiana–Mississippi border. The northern eyewall was still reported to be very intense by WSR-88D radar data and the intensity was estimated to be near 121 mph. Even an hour later and far from the center, Dauphin Island, AL reported sustained winds of 76 mph with a gust to 102 mph, Mobile reported a gust to 83 mph, and Pensacola, FL reported a gust of 69 mph.

Katrina continued to weaken as it moved north northeastward during the remainder of the day. It was still at hurricane strength 100 miles inland near Laurel, MS. The storm was reduced to tropical storm status by 19:00 CDT when the storm was 30 miles northwest of Meridian, MS, and became a tropical depression near Clarkesville, TN on August 30.

Chapter 5

I n what he would remember as a "somewhat infamous" email, Gary
Marchand, the CEO of Memorial Hospital of Gulfport, issued a
Katrina alert to all hospital employees on Friday morning, August 26, 2005.
Not reading but recalling the email's essence later, he communicated, "Not
to alarm anybody, but we know it's out there. We're watching it. Keep your
wits about you. If things change we'll be in touch." The storm would hit
Apalachicola, Florida, he affirmed.

That was Friday.

The scenario changed rapidly—not so much that day and the next in
Gulfport but, certainly, in relation to what weather forecasters predicted.
On Saturday the hospital's executive management team assembled in the
executive dining room, the most spacious, secure, interior room they all
knew as "command central" for any emergency situation. Sipping coffee
and reviewing what they already knew about the approaching storm, they
collected new information from television broadcasts and checked in via
telephone with Harrison County civil defense officials.

"Normally and historically, when any storm starts to move across the Gulf,
if the weather people say it's going to hit Florida, it's going to hit Alabama, it's
going to hit Mississippi, and then they say it's going to hit New Orleans—
those storms rarely track back," Vice President for Administrative Services
Lawrence Henderson reminded his colleagues. But they already knew that
the storm in the Gulf of Mexico would not follow the norm. "That thing
has an eye presentation that's just ominous," Marchand observed.

Reaching agreement to go on hurricane alert at 5 pm Sunday took little time. Picking the day and time set into motion a specific and detailed series of tasks to inform and cue into action two teams of approximately 620 individuals each: the hurricane team and the relief team. "Everybody on the hurricane team must arrive with pillows, blankets, snacks, clothes for three days, ready to work . . . We do not allow, as a matter of policy, our employees to bring their families with them—I'm sure it would probably violate HIPAA (The Health Insurance Portability and Accountability Act of 1996) and they tend to consume your power, your food, your water, and clog up the hallways," Marchand gently but firmly reminded.

Practitioners from each medical or surgical specialty also attended, covering every area: obstetrics, urology, general surgery, interventional cardiology, cardiovascular and open heart surgery, anesthesia, and trauma. The hospital had sufficient medical staff resources to cover three to five days.

Memorial Hospital of Gulfport occupies the entirety of more than four city blocks, facing the Gulf of Mexico at 4500 13th Street back to 15th Street and stretching from Broad Avenue on the west to 44th Avenue on the east. The hospital sits some three blocks north of the railroad tracks and about two miles west of the Harrison County Courthouse and Emergency Operations Center (EOC). Under joint ownership of city and county, the facility—one of Mississippi's most comprehensive healthcare systems—provides 445 beds, including an inpatient rehabilitation unit, a behavioral health facility, and more than 50 Memorial Physician Clinics. The campus features four primary buildings, the Main, East, and West Towers plus the Medical Office Building and a large parking garage as well as outlying structures for ancillary services. The structure—built of reddish brick, black glass, and shiny steel—commands both community pride and loyalty.

Actions the hurricane team takes before the storm—or for any emergency situation—are detailed in the hospital's "all hazards" emergency preparedness and response plan, by law on file with the licensing authority, Mississippi State Department of Health. The document gets regular, routine attention throughout any year, and every activation usually results in improvements based on new experiences. Particular areas of emphasis cover communications, resources and assets, safety and security, staffing, utilities, and clinical activities. Every aspect of the plan aims to provide for

Memorial patients exactly the type and level of services they need.

"We update the plan every year or before every hurricane, whichever happens more frequently," said Diane Gallagher, vice president for marketing and planning. "We can go back through the years and identify changes based on any particular storm and/or developing protocols. We've had *significant* changes since Camille. And for every event, not just for every storm but also for every scare, we look back and try to determine what we could have done differently that would have made the outcome better."

In preparation and in providing for their own comfort, individual employees focused on what was most important to them. "I brought clothes for a week, my sentimental jewelry, and a couple pairs of shoes," said the nursing services/behavioral health manager. "I also brought a toy my dad played with in Sweden when he was a little boy—and the Swedish Bible I am supposed to take with me everywhere I go, according to my mother." A physician, who describes himself as a music student and intermediate player, brought all his music and his violin. Another brought "insurance papers, house papers, and my Taladega tickets because I was going to the races, come hell or high water! And we got both of those."

Hospital Administrator Marchand spoke for the whole team: "This storm was a little more than we planned on."

As is customary, physicians released hospital patients who were able to go and could likely be safe at home. They also considered whether to admit people with special needs. By 5 pm Sunday the huge hospital complex chugged along as only a big operation can—everything apparently calm and running smoothly. Nervous anticipation provided the undercurrent, but staff and patients did not talk much about the storm.

"None of us realized how bad it was out there," Gallagher remembered. "We knew it was pretty bad, but we certainly did not realize it would be as devastating as it was. We did not realize the miracle we were until a long time after."

Winds picked up about 2 am Monday. "In the wee hours of the morning," Marchand said, "we knew. It was scary to pull up the Internet sites. When the eyewall came into view on the radar screen, it would just scare you to death to look at that thing. It's just coming right at you, and you know it. And then you try to marry the radar perception of what's still out there with everything that's currently busting lose around you and—personally, I found it difficult to connect with all that."

Winds roared, rain pounded, lightning cracked, and windows began to explode. Engineers scrambled to keep the three sets of front doors closed. Four or five men struggled to secure the front entrance. From there, they looked up to see huge chunks of glass panels fly off the West Tower. "Yep! There it is," one called out. Pieces of the roof ripped off, a big air conditioner unit atop the fifth story blew, crashing into what had been the old medical records department—unoccupied because the space was being remodeled. Within minutes of the first window's popping, nurses and aides rushed to remove patients to interior hallways.

"All we had that we were certain of was each other," Gallagher recalled. "We weren't certain of our families. We weren't certain of our homes. We had each other, and we had bonded—not just the management team; I'm talking the whole team—there were no doctors or nurses or housekeepers or food people or administration. We were all just right here together in our own little world. Dependent on each other."

From early morning Monday, Code One alerts announced every new breach, every instance of hurricane damage. "And that was like increasing the anxiety for everybody inside the building," Gallagher said. "So we—we know it's going on. It's bad. Shut off the announcements because we know things are being stripped off the buildings and flying around."

As if Katrina had smashed Pandora's Box onto the coastal feet of Mississippi, demolition dinged and donged every surface in her path. Despite engineering's best efforts, the glass front doors blew out, causing the next set of doors to come off their hinges and allowing the third set of wooden doors to rattle. Flapping and banging indicated total havoc; instead of calling alerts for individual rooms or spaces, internal communications informed of whole hospital wings having been damaged.

"The West Tower," an engineer identified. "There's so much glass, and it's getting hit the worst." Windows vibrated, back and forth; then a loud pop! And the windows had been sucked out. Security and engineering helped pull patients out of the rooms and tied the doors shut to prevent people from being dragged into the storm's fury. They used towels and blankets under the doors to keep rainwater out and tied pillow cases around patients—anything to protect them. "Pardon the expression, but all hell's breaking loose! We're putting out one fire and running to another," another employee yelled. About 40 feet of one wall pulled away from the structure and crashed below.

"I never would have believed the storm could do what it did," one engineer said. Air conditioning units being blown off the building and windows being blown out—these windows are supposed to hold 300 mile-an-hour wind!"

People's screams pierced their cacophonous world. Another loud ka-boom! And light fixtures sucked out. Ceiling tiles undulated, snake-like. "Are we going to die?" implored a teenage girl. "No, we're not going to die," affirmed a nearby nurse. "I refuse to die at work. I am not going to die here! Just keep down."

Throughout the 12 hours Katrina battered Gulfport, Memorial Hospital's command central remained vigilant. Monitoring media and phones as long as they had external communications, the lead managers also walked around. They needed to see actions taking place throughout the huge complex and that patients were safe and undisturbed despite the demonic sounds snapping all around them. They needed to breathe in the security of their own offices, and from those management-by-walking-around excursions, they could occasionally also see outdoors. Katrina was a daytime storm. She began to arrive before dawn, or what would have been dawn, and landed along Mississippi's Gulf Coast near Waveland in adjacent Hancock County about 10 am. But her monstrous size—no less than a 250-mile concentration of pure evil strength, swirling and hurling—put her fury atop Harrison, Hancock, and Jackson Counties for that whole Monday and into the night. Her third landfall near the Louisiana-Mississippi border that morning with 120 MPH (190 km/h) sustained winds, still at Category 3, seemed to indicate her acceptance of Mississippi as "the hospitality state." Would she ever let up and move on?

Marchand thought about his last phone call—his wife's telling him a pine tree branch was sticking into the second floor sheetrock. His secretary heard her own child: "Mom, the roof is lifting off and onto the house!" Then the connection died. From another colleague, "Daddy! Water's coming into the house!" Then, silence.

Another manager on the relief team and, therefore, at home during the tumult, called to tell colleagues he and his family were going into the attic because water was coming into the house, but the roof was gone; so they decided to go outside to seek shelter.

"We didn't talk about it," Marchand said. "Clearly, none of us believed they could possibly live through that. I felt guilty because we could so easily dismiss that thought during our own distress, but we didn't talk about it. There's no way they could make it out alive."

As Katrina pounded, the command team lost touch with the outside world. Television signals scrambled and then disappeared. Telephones quit working except for intermittent connectivity. "This should have come and gone already," Marchand said. He tried again to reach somebody at the EOC. Dr. Robert Travnicek answered. "How much longer?" Marchand shouted above the wind's roar. "Another two or three hours," the public health physician said. "Tell me what your situation is regarding patients now."

Loss of communications, of any reference beyond their own stronghold —that's what most frustrated every one of the health care professionals. Information Technology (IT) workers kept the internal computer system working from their space on the East Tower's fifth floor. Even when part of their wall ripped off, they hung tight. Marchand had to go see for himself. Bedraggled in t-shirts and short shorts, they sweated from computer-generated heat, knowing they stood mere inches from electricity under the floor and also knowing they were mere yards from a concrete stairway where, if things got really bad, they could just pop out to safety in the stairwell. "Let us stay," they implored the CEO. Eventually, even their heroic efforts could not save the system.

"This was in the afternoon," Marchand remembered. "Part of the fifth floor wall we lost exposed our communication switches to the weather. You could open a door to the room, but the storm tends to try to suck you out and Burt was trying to get into that outside! 'Well, if we can get in there, I can cover it with Visqueen, and the time switches,' he said before I yelled, 'Nobody's going into that room!' And, surely enough, nobody went in there. That's when I remember thinking, 'How long is this going on?' It had to have been four or five o'clock in the afternoon."

Chief Nurse and Patient Care Vice President Jennifer Dumal knew the computerized clinical systems had failed. "It didn't seem to abate. It just stayed the same."

"It was bad, bad, bad," Marchand recalled. "It just never stopped. You could definitely hear the wind, and we could see from our office windows. From the sixth floor, the outside is just water. Some of the water pipes broke, and you'd think it would be total chaos. But it was calm. It was work. Matter-of-fact. The only thing we could do was check out what was going on. Throughout the hospital, in the OR, people were just working. Patients were fine; they had a family member with them, they were with us, and they were calm. As chaotic as this sounds—with the windows breaking and the sounds of the wind and rain, we had it under control. And the *appearance* was that we had it under control. Everybody was just working."

Chapter 6

Flashback four years to October 2001, directly across North State Street from the University of Mississippi Medical Center in Jackson: Mississippi State Board of Health's routine quarterly meeting. The policy-making body for Mississippi's public health system played its political hand; after a closed executive session and no small amount of nervous tension among Department of Health employees and public health advocates for several special interest areas, State Health Officer Ed Thompson announced his retirement—to become effective 15 months later.

This event foreshadowed a political disturbance that would enshroud Mississippi public health for half a decade. In a similar manner to the 2005 hurricane that would adversely affect Mississippians for years to come, the public health mess would ferment, fester, and finally burst at the hands of the state legislature.

By July 2002, the Board of Health had hired a headhunter, supervised a nationwide recruitment campaign, interviewed four candidates, and selected Thompson's successor, Brian Amy. For the first time in history, the Mississippi State Department of Health's executive officer would not be native to Mississippi.

The board-appointed Louisiana man would be state health officer and executive director of the public health agency, historically free from political bondage because board members served six-year, staggered terms appointed by non-successive governors. State law changed composition of the board during Governor William Winter's administration so that consumers as well as providers could represent the public health interests and so that no governor would have appointed a majority of the board's twelve members. That changed dramatically, however, when Mississippi in 1986 approved gubernatorial succession and a sitting governor could appoint a board majority.

No longer a body of physicians only, in 2002 the Board claimed five physicians, a chiropractor, a pharmacist, a dentist, two nurses—one a nursing home administrator, a long-term care homes owner, and one consumer representative. One of those members had filed a lawsuit against the Department of Health before his appointment to the board and then filed yet another suit against the state health officer and two employees responsible for health facilities licensure and regulation after he became a board member. That same member reportedly bragged that he "bought" his own seat and also the seat of a member reappointed in 2002—cost: $25,000 each in political contributions to then-Governor Ronnie Musgrove.

Within nine months of Amy's appointment and at the behest of a long-time public health employee, retired, the Mississippi Legislature's Performance Evaluation and Expenditure Review Committee (PEER) began a full-fledged investigation into the department's management and the board's reported "conflicts of interest, deception, and lack of attention to health protection." The PEER petitioner cited political shenanigans and questioned the professional, political, and personal affiliations of the state health officer, rumored to have greased his way into the position with a sizeable monetary contribution to the governor's political fund. The requestor alleged that board members engaged in political interference in agency operations, particularly the regulatory functions of health facilities licensure and environmental health science, and that the new agency director stimulated illogical personnel practices along with cover-up and inappropriate use of state funds and property.

Just as Hurricane Katrina formed from nothing and became a monster storm in 2005, the power-mongers who comprised the State Board of Health and executive management team of the State Department of Health in those early years of the new century's first decade also gained girth and grit. Opposite that faction, in August 2005, Dr. Robert Travnicek, an unsuccessful candidate for the agency's directorship in 2002, found himself under fire and at the mercy of both that disturbed state health officer and a disagreeable State Board of Health.

Robert G. Travnicek, MD, MPH, moved to Mississippi in 1990. Freshly graduated from Harvard University with a master's of public health degree and having audited courses in the Kennedy School of Government, the Business School, and the Medical School, Travnicek turned his focus from 22 years of private family medicine practice in Nebraska to a new career on the Gulf Coast.

This near-the-turn-of-the-century transition to Mississippi marked Travnicek's second going to the Deep South. A seminal year in the early 1960s had put the young US Public Health Service physician in the state during Freedom Summer, at the height of the civil rights movement. That's when he developed a long-standing professional friendship with another young doctor who in 1973 would become Mississippi's fifth State Health Officer, Alton B. Cobb, MD, MPH. By 1990, Cobb had become one of the nation's longest-serving and most highly respected state public health directors. When Travnicek tired of private practice and connected with Cobb about returning to public health and Mississippi, the senior physician challenged him to first earn an MPH degree. With the credibility of that accomplishment from Harvard, one of the country's most highly respected schools of public health, Travnicek landed the district health officer post for Mississippi's six-county Coastal Plains Public Health District. Cobb predicted that the seasoned physician with a fresh approach to public health practice would fit well with the more progressive population of Hancock, Harrison, and Jackson counties and also serve well the interests of Pearl River, Stone, and George counties.

As district health officer, Travnicek was responsible for medical supervision of all Mississippi's traditional public health programs: maternal and child health, chronic diseases, communicable diseases, epidemiology, and environmental services—food safety, water supplies, sewage disposal, radiological health, and health facilities licensure and certification. During his first 12 years on the job, Mississippi State Department of Health (MSDH) both changed and grew. The agency's mission clarified over time to promise protection and promotion of the population's health as a whole rather than to focus on treatment of individuals. Mirroring a nationwide trend to strengthen public health, MSDH committed to balancing three core government public health functions: assessment of communities' health status and resources, development of health policy and recommendation of programs to meet health needs, and assurance of the availability of essential, high-quality, effective services. This required public health workers, especially those at the local level, to involve themselves more strategically with private practitioners and non-profit agencies, including community health centers and government public health agencies. Local public health workers were to involve themselves more directly with people beyond the health department. The agency's central office put efforts toward program planning and policy guidance plus administrative and technical support for staff in the districts and counties.

The model well suited Travnicek, who developed team-player-oriented relationships with employees in the various disciplines that comprised public health workers: nurses, environmental health inspectors, aides, clerks, social workers, and other physicians. At central office meetings in Jackson, he sometimes seemed disconnected, even aloof, but remained keenly aware of the individuals and schisms surrounding him. In 2005 hurricane season, he was a public health physician in upheaval not of his own making but caught in a quagmire of natural disaster, local and state politics, and moral determination.

Travnicek talked in late 2005 about the change, concerned about what he saw as a degradation of the physician's role: "It's a state system but—ideally, it functions on a local level as basically independent health units. The department had undergone a redefinition of that, and that's being generous, but a redefinition of that where we are more totally centralized in the past three years, where no decision can be made locally, including answering the phone! If I get a press call, I have to send it to Jackson. I used to do hundreds of TV appearances, but I haven't done a single TV appearance in the past three years. It's been a department that is increasingly difficult to be professional in, and I think that's the best way I can put that. Morale in the department—it's gone from bad to worse on a virtual daily basis."

Self-aware as few people are, Travnicek focused attention as a public health doc on the whole population of about half a million people pre-Katrina: "I'm the kind of guy who is basically a lifelong Democrat but realized, like Rodney King, that we all have to get along; otherwise, there's going to be civil war here. So I tend to bridge. I tend to do a lot of things which on the surface have absolutely nothing to do with public health. I go through a lot of calls, and I know a lot of people who have money, and I'm not a guy who's a naysayer running around wringing my hands. I realize that if there's social unrest or plagues—if there's a general plague that would also affect people who have millions, probably to a lesser extent but affecting everybody. . ."

Hurricane Katrina would affect every strata of society, all the population, but she was not Travnicek's first disaster response.

"I've been in the Harrison County Emergency Operations Center since 1990," he said. "For every storm, every miss, every near miss, through three emergency directors including Wade Guice, who's the dean of hurricanologists in this whole country. He's the one that brought the Coast through Camille and recognized as being the dean guru. I mean, when Hugo hit South Carolina, they sent Wade over there. And he was a good, personal friend of mine; I did the eulogy at his funeral, at his request. So there's a lot I know about this. I've been in that EOC as a medical guy, even though originally he didn't even see the need."

Strategically positioning himself to both learn and teach, Travnicek devoted hours on end beyond hurricane season with Guice in the EOC. "He realized I am mightily interested in this. This is a unique part, the environmental aspect, because of the environmental sensitivity here: wetlands, oyster reefs, shrimp, general pollution of the whole environment and of the Gulf of Mexico. It's fundamental public health.

"But you can't really enter any space, much less an EOC, relate, and lead the first time you meet somebody. You have to be here awhile, and you have to know what you're doing to really make the thing work, so for the past 15 years, I would go when they activated the EOC. Steve Delahousey, who is my counterpart for ESF-8, and I always went there. We go, and we start meeting, and we start talking to hospitals and nursing homes. We hold separate meetings because the general meetings tend to be too big, and we figure out who needs evacuation and all that sort of thing.

"When we sit down, we've got two projectors, we've got the National Hurricane Program's computer-based storm tracking and decision-support tool called HURREVAC; we've got all the models and direct links with the National Weather Service. We're pros at this. Wade Guice set it up that way."

For three days before hurricane watchers expected the storm's landfall, Travnicek divided time between directing the public health district and working in the EOC. On Friday, he acted to "save the staff, save the buildings, save all our pharmaceuticals—board everything up, and tell them to get the hell out of here, to wait until the storm comes, and then come back. This is standard operating procedure."

Even though he and Delahousey had handled the situation for a worst case scenario, everybody "kind of thought the storm was going to go someplace else. We've had several near misses. Ivan was a near miss. Dennis was a near miss. Those all went off; it looked like this thing was wobbling. We're watching all the weather services, NOAA, and all the media—each one of them is assuming the storm's going here; no, the probability is the storm's going here; and many of the models were showing it going to Louisiana marsh."

No doubt, if only because of the girth and the physical magnitude of Katrina, the Gulf Coast would get some weather. Sunday afternoon, both Lora Travnicek, a nurse at Garden Park Hospital, north of I-10 near Highway 49, and her husband went to their workplaces.

"We don't have any kids at home," he said. "So we just locked the house up, bolted and boarded it up, and I took her to work and came to the EOC."

Travnicek had no worry about his property. Before buying on the Gulf Coast, he consulted Wade Guice, who emphasized that a "safe" property would be at least 30 feet above sea level and a mile back from the water. More than a mile and at 47 feet, he felt secure.

In final preparation for Katrina, the Travniceks closed up, filled both vehicles with gasoline, made sure the generator was in place and ready, and went to work. All action Sunday occurred as the veteran public health doctor expected—until Sunday night.

About eight o'clock that evening, both Delahousey and Travnicek realized anew the storm's potential to slam directly into Mississippi.

"About that time, a whole bunch of people we've never seen came in," he would remember. "You've got to have clearance to get into the EOC."

"Who are they and why are they here?" Delahousey asked.

"This isn't a place for people who don't have a job—not for just walking around. I don't care if you're with CBS News," Travnicek responded.

One of the newcomers assumed the spokesman role. Standing in the front corner with county supervisors, he talked about potential mass casualties and other ills the storm could bring. The speaker was Mike Beeman, just arrived to represent FEMA before the storm made landfall. The new folks who arrived about the same time had evacuated the Hancock County EOC at FEMA's directive. The few who chose not to leave then would have no option at the height of the storm: hang on, swim, or drown.

In similar dire straits were the individuals who telephoned throughout the night, despite officials' having begged them to leave. Callers clamored for somebody, anybody, to come rescue them.

"I don't dwell on it," Travnicek said later. "We'd tried all Saturday and all Sunday to get people to evacuate. I don't think our messages hit hard, and I feel responsible because I didn't get—my department wouldn't let me—on television. When your doctor tells you to leave, you're more likely to listen than to somebody you don't know. Spraggins did a good job, but he was new; nobody knew to listen to him. Many of them saved their own lives by clinging to trees, but a lot of people did not take the messages seriously. So we're fielding calls from people who were in houses and the water was coming up. The water's coming up, and they're dying. I told them to save their cell phone, try to keep their cell phone dry, get as high as they could and, if they were God-fearing people, they were. . . their life was in the hands of the Lord."

Chapter 7

August: hottest month of the year in Deepest South Mississippi. Thirty-one seemingly interminable days with little, if any, relief from blazing sun, unrelenting heat, and humidity. In 2005, the month of August became calendar home to five named storms: Tropical Storms Harvey, Jose, and Lee and Hurricanes Irene and Katrina.

Atlantic hurricane season started that year with Tropical Storm Arlene—affecting Honduras, the Caymans, and Cuba before landing near Pensacola, Florida, and costing more than $11 million. By year's end, the season would encompass the entire alphabet (save any beginning with the letters Q, U, X, Y and Z), with 15 storms reaching hurricane status. Hurricane Wilma landed in Cozumel, hit the Yucatan coast, and travelled across southern Florida to claim 35 United States citizens and cause $28.8 billion in damages. A rare year with more than 21 tropical storms, 2005 also saw tempests named by the Greek alphabet: Tropical Storm Alpha, Hurricane Beta, Tropical Storm Gamma, Tropical Storm Delta, Hurricane Epsilon, and Tropical Storm Zeta would follow. The season's final formation of Zeta began December 29, four weeks after the hurricane season officially closed, and finally dissipated January 6, 2006.

On August 29, Hurricane Katrina picked Mississippi for her third, and final, landfall, ultimately claiming 1,836 souls and costing upwards of $115 billion in damages throughout Louisiana, Mississippi, and Alabama as she traveled northward through the central United States to the Great Lakes. A cold front got her there, and she died in southeastern Canada.

Along the way, she transformed the season and temporarily changed the mood of Mississippi people: hot to cold, happy to haunted, thriving to wasted.

Many people in South Mississippi went to bed Saturday night, August 27, vaguely aware that a storm was in the Gulf of Mexico but more concerned with enjoying family and friends in those final days of summer. Cooking out, going to the beach or boating in Mississippi Sound, shopping for back-to-school. . . Who knew that by 7 am Sunday, Katrina would swell into a full-blown Category 5 hurricane? Within a day-and-a-half, she would slam into Mississippi's coastline, the third strongest hurricane on record to make landfall on the United States. Only Hurricane Camille in 1969 and the 1935 Labor Day Hurricane boasted more power.

From the previous Wednesday, many who lived along the Gulf Coast, especially people of Jackson, Harrison, and Hancock Counties, called Jackson metropolitan area hotels for safety from the storm. Newspapers reported Sunday that swamped hotels were "sold out" and had waiting lists. A dozen shelters in the area had opened, nearly 600 Tulane University students and athletes had evacuated by pre-arranged plan from New Orleans to Jackson State University, and the City of New Orleans stopped its southbound trek from Chicago in Memphis, 395 miles north of its destination.

Mississippi Emergency Management Agency (MEMA) officials predicted a "tremendously dangerous storm" but feared that many people would not evacuate. MEMA Executive Director Robert Latham expected the storm not only to arrive but also to move slowly and not clear south Mississippi until after midnight Monday. South Florida continued to hurt from Katrina's earliest landfall: with temperatures in the 90s, 733,000 homes and businesses still had no power.

Even though Katrina would not slam into Waveland, Mississippi, until mid-morning Monday, that day's edition of *The Clarion-Ledger* proclaimed "UNIMAGINABLE" across its front page because editors knew the Category 5 storm had aimed directly to hit Mississippi's Gulf Coast. During and until well after the press run, Coast emergency workers continued warning their neighbors to evacuate. From Greg Doyle, AMR and Harrison County spokesman from inside the EOC in Gulfport: "Nothing is a good reason to stay. Police, fire, emergency medical technicians, and the National Guard are on the ground, but people who stay behind cannot look for any help during this monster storm. It's too dangerous."

MEMA said anybody living south of I-20 in a mobile home should move to a stronger shelter because Hattiesburg, McComb, and other municipalities miles inland could see 100-mile-per-hour winds. Pike County officials prepared for many injuries, and Meridian retailers remarked that their customers seemed to know "this one is different." As Katrina barreled northward that Sunday, New Orleans expected 160-mile-an-hour winds and feared that air currents could push Gulf waters into Lake Pontchartrain. The Superdome already housed 35,000 New Orleanians. Entergy crews from Arkansas set up at the fairgrounds in Jackson, and Texas designated 90 members of search and rescue teams to help Louisiana. Mobile braced for flooding as many roads in the Florida Panhandle already were under water.

At MEMA's cramped headquarters on Riverside Drive in Jackson, Director Latham hosted the FEMA team that had arrived on Saturday. This was not their first hurricane watch for the season.

"About a month, a month-and-a-half before Katrina, we had Dennis, kind of a menace storm," Latham recalled, "and we had the FEMA team here for Dennis. In 2004, four hurricanes hit Florida; our FEMA support normally comes out of Region IV, which is Atlanta; our Federal Coordinating Officer (FCO) and all of his staff would come from Region IV. The problem is they were all still tied up in Florida from the four hurricanes in 2004; so when we got Dennis, our backup region was Region IX, headquartered in California. Part of that team deployed here for Dennis, and the FCO was a guy by the name of Bill Carwile.

"Bill was one of the older, more experienced federal coordinating officers for FEMA. He had done the first World Trade Center, Oklahoma City, the Loma Prieta Earthquake, and a lot of disasters that were in the Pacific. I met Bill during Hurricane Dennis. For Katrina, they deployed basically the same team back to Mississippi because *our* team was still in Florida.

"When it was obvious that this was going to be catastrophic, Bill came to me and said, 'I want to try something. I've got this idea of taking the Incident Command System (ICS)—structured and scalable, designed with planning, operation, logistics, financial, with an incident commander—the whole structure.' He said, 'I've got this concept that works good on fire-ground operations, on a single incident; if it's a fire, the fire chief is in charge; if it's an act of terrorism, law enforcement would take the lead because it's a crime scene. But something like this is going to be so big and require so much effort and coordination of resources, I want to try this unified command concept.' He drew it out for me—it was the ICS structure but was basically pairing the federal and state workers. For instance, for operations, you'd have two—you'd have an operations section chief from the federal and from the state; same thing with logistics, same thing in financial, public information, external affairs, same thing at the unified command level or the command general staff level. I would pair with Bill. And I said, 'Well, I'm not sure there's any way we can deal with this unless it *is* a unified effort.' So we did it.

"And that's how the unified command structure—we were able to use it here. We in Mississippi are the model. We're the first disaster event in which it was ever used. Bill actually had tried to do it in 2004 in Florida when he was the FCO there, but he said, 'I just never could get it going in Florida. I just never could get it pulled together.' So when he approached me about it, and we did it, it became the model. That's how it's taught now: unified command."

Latham and fellow emergency preparedness officials had been teaching the concept of incident management for a good 15 years before Katrina. But between 9/11 and Katrina, the method changed. Within a couple years before August 2005, the training included "all the people who are going to be working together," Latham said. "It was actually a bioterrorism scenario that we did in those workshops, and we did one in every county, 82 of them. We brought together all the people in that community—public health, law enforcement, fire, EMS, emergency management, elected officials— everybody. And we had them sit around a table together, everybody; gave them a scenario; and they walked through it. The product that came out of all those workshops—and we finished them about six months before Katrina—was a template of how an ICS structure would look. Here it is. All you have to do is put the names in there for who is going to fill what role. We had actually completed that about six months before Katrina."

That intense focus started during the administration of Mississippi Governor Ronnie Musgrove, who signed Executive Order 851 on October 16, 2001, to establish the National Interagency Incident Management

System (NIIMS) as the State Incident Command System to be used during emergency operations.

"In those days," Latham mentioned, "it was a two-I'ed NIIMS and based on fire service operations. Later the system was changed to a one-I'ed NIMS, which is the National Incident Management System. The name and the acronym changed; it evolved. And Mississippi was the sixth state in the nation to adopt the system."

Those days—immediately after the terrorist attacks of 9/11, are the first time lives were lost on American soil since the bombing of Pearl Harbor during World War II—and the ensuing days and weeks now known in law enforcement circles as "Amerithrax." Department of Justice, FBI, and US Postal Service spent years investigating the appearance of anthrax-laced letters, sent first to Senator Patrick Leahy on Capitol Hill and later to numerous other suspected recipients; the toxic letters killed five citizens and sickened 17. The biological attacks combined to become the worst in the country's history. Being prepared for bioterrorism took on new meaning.

Just 11 days after September 11, 2001, the White House appointed a new director of Homeland Security "to safeguard the country against terrorism and respond to any future attacks." Following the presidential lead, in November 2002 Congress created the Department of Homeland Security as a "stand-alone, Cabinet-level department to further coordinate and unify national homeland security efforts."

In the wake of terrorist attacks and the anthrax threats that unfolded that year in Washington and New York, Mississippians adopted a watchful attitude towards events they previously would not have questioned. The public paid closer attention to any and every white substance and mysterious powder found—powders which turned out to be sugar, baby powder, soap flakes, and a number of other inoffensive substances.

Neither the federal nor state government could afford to disregard threats or potential threats to its people. Beyond bioterrorism, officials turned new attention and allocated new money to being prepared also for contagious disease outbreaks, chemical emergencies, radiation events, and mass casualties. The Centers for Disease Control and Prevention became the conduit for hundreds of millions of dollars devoted to preparing for and responding to public health emergencies. For its part, between 2003 and 2006, Mississippi State Department of Health garnered more than $44 million for preparedness and response to bioterrorism, emerging infectious diseases, and other public health emergencies. The money paid for people, planning, training, supplies, equipment, and communications.

Change became the constant.

Particularly important to public health preparedness: a soon-to-be-revealed critical insight to poor management decisions within the statewide public health agency. Just weeks after Hurricane Katrina's landfall, Mississippi Legislature's Joint Committee on Performance Evaluation and Expenditure Review (PEER) published its report of management turbulence within the Mississippi State Department of Health. PEER's review of the department from October 2002 through August 2004 cited the agency for having implemented four organizational changes in its structure within just 23 months—with two of those "major revisions" from the structure in place for six years and accomplished without "formal approval" of its own policy-making board. PEER said that "precludes departmental personnel from developing the working relationships necessary to accomplish the organization's mission." PEER also found that the management team "changed the channels of communication for staff members without clearly stating the intent of or goal for the changes and without documenting the desired communication procedures in formal, written policies or staff memoranda. The management team has also restricted traditional professional channels of communication and relationships with external information sources and with public health providers, a situation that could affect the staff's ability to promote and protect health."

Despite that environment, MSDH claimed success for drawing more than 200 Mississippi physicians to a "first-ever training on basic disaster life support" in June 2004, an event jointly sponsored with Mississippi State Medical Association. Jim Craig, who advanced from his position as emergency planning and response director to director of the agency's broader and more inclusive health protection unit in March 2004, said the course would "help Mississippi physicians prepare for the unexpected. Physicians and healthcare professionals must be ready—they may be the first to see the patient."

Craig's office was "responsible for preparedness, plans, and responses for public health and agency needs created by a natural or man-made event" and managed Mississippi's access to such federal resources as CDC's Strategic National Stockpile and hospital preparedness activities under Health Resources and Services Administration (HRSA). The office coordinated responsibilities with and through MEMA, according to the Mississippi Emergency Response Plan.

As part of that plan, both Craig and MEMA Director Robert Latham participated in the 2004 disaster training exercise that would become known as the "preview" into Hurricane Katrina. Federal, state, and local planners designed "Hurricane Pam" to help address response issues that would require collective grappling in the wake of a catastrophic hurricane; they projected the event onto New Orleans, Louisiana, and involved up to 350 participants in a series of structured sessions. They aimed to gather and prepare planners and decision-makers from all levels in developing their integrated duties under the National Response Plan (NRP), finalized in December 2004 and formally announced in January 2005. News accounts touted the NRP as a "comprehensive, all-hazards tool for domestic incident management across the spectrum of prevention, preparedness, response, and recovery," said Secretary of Homeland Security Tom Ridge. "The complex and emerging threats of the 21st century demand this synchronized and coordinated plan in order to adequately protect our nation and its citizens."

During the months between fictional Hurricane Pam and real-life Hurricane Katrina, Mississippi's emergency management and public health agencies coordinated efforts to assure compliance with the NRP. The Pam exercise put together planners and operational personnel to cooperatively figure out what they would need to do and expect. In and after that drill, Latham and Craig identified and then tried to fill some gaps—for example, before the storm, Latham sought public health help to support special medical needs shelters.

Craig recalled their approach: "Until Katrina we didn't have that mission; there were no special medical needs shelters. There were shelters that Human Services and Red Cross did the best they could to take care of folks. But it was recognized there was a gap in taking care of, for example, hospice patients who were at the end-stages of their life, people with other medical requirements, or who had been on home health services. So Robert actually came to visit to see if there was an opportunity for MSDH to help. Knowing that there were many of our nurses who needed to be able to assist in some of our medical pieces as well as to get a safety net of our public health activities back up early—we'd never really played a role in the sheltering piece. But because they were very concerned about the special medical needs piece, and given Red Cross's command at that point in time, they asked us if we would take that additional mission."

Harrison County ESF-8 coordinators knew that some 25,000 non-institutionalized special needs individuals lived in Hancock and Harrison Counties and expected that only about 10,000 would choose to or could evacuate before the storm. Robert Travnicek, MD, MPH, and Steve Delahousey, RN, EMT-P, worried about every one of them.

Chapter 8

Fewer than 20 miles west of the Harrison County EOC, Hancock County Emergency Management Director Brian "Hooty" Adam took the call from National Weather Service–Slidell late Saturday night or early Sunday morning. "We're going to be ground zero."

From the Thursday before and for the next 72-plus hours, Adam had been and would continue to be in "get ready, get more serious" mode. Preparations for the late August 2005 hurricane ramped on Friday, August 26, and people representing the emergency support functions for disaster management poured into the Hancock County Emergency Operations Center on Highway 90 in Bay Saint Louis.

Locals, transportation, communications, public works, health department, firemen, and law enforcement, Adam would recall. He also said "a gentleman named Eric Gentry" would come in as FEMA coordinator; the National Guard came; and state emergency management—including Director Robert Latham—also appeared. They all came before the storm.

"Hancock County's 485 square miles with 195 square miles in low-lying areas," Adam reminded them. "Those areas are susceptible to flooding at high tide and with a good southeast wind. We have a lot of homes, trailers, different things in that area. It looks like it's going to be pretty serious."

Friday's discussions still studied the unknown. Models showed the storm could hit anywhere from West Florida to Mexico. Supervisors, mayors of the two incorporated towns, Waveland and Bay Saint Louis, fire- and police chiefs: nobody really knew, but all acknowledged the seriousness of the obscure. Saturday's all-hands-on meeting drew more than just elected officials; individuals representing the hospital and AMR ambulance service appeared, and all county workers were on notice.

An unlikely structure served as the Hancock County EOC. An old bowling alley about four or five blocks from the beach primarily furnished justice court offices and courtrooms and also provided space for emergency management, kitchen, a food pantry and food storage area with a big walk-in cooler, as well as space for the Red Cross. Near the middle and on the eastern side of the concrete-block building were the radio and "war" rooms.

"We're getting squeezed into the eye," noted Chancery Clerk and County Administrator Tim Keller after the Weather Service's call. "We need to get the jail trustees to cover the file cabinets and start securing the courthouse and other county buildings. Rain's coming."

Common in many counties who count on few to do the work for many, Keller's post also positioned him as public information officer.

The Hancock team met again Sunday morning and agreed with Adam to go straight to mandatory evacuation for everything in low-lying areas and all south of Interstate 10. Even as the officials threw out public service announcements via Mississippi Coast and New Orleans mass media, people claimed their homes had been safe from Camille.

"This is a different storm from Camille," Adam emphasized. "It's going to hit us on the eastern side, whereas Camille hit us on the western side of the eyewall. We had the eye come *over* us in Camille, but we were on the western side of the eye, not the eastern or more dangerous side. When the water's on the eastern side of the eyewall, it's getting *pushed* inland. . ."

National Oceanic and Atmospheric Administration describes the eyewall as an "organized band of clouds that immediately surround the center, or eye, of a hurricane. The most intense winds and rainfall occur near the eyewall." Scientists would study Katrina and determine her to have been a "double eyewall storm," and years after, NOAA Scientist Jim Kossin would use her data to develop a new model that can help predict a cyclone's life cycle. "Hurricanes usually strengthen and grow gradually over time, but eyewall replacement cycles can cause very sudden changes in size and intensity," he said. Hurricane Katrina weakened as she moved northward over the Gulf of Mexico "but grew in size because of an eyewall replacement cycle, and the huge wind field led to an enormous storm surge that devastated the Gulf Coast."

Adam and his colleagues in the EOC that hot August weekend relied on what they knew at that time and from their own experience. They knew any northward-moving hurricane's northeast quadrant could be deadly.

Sometime Sunday—neither Adam nor Keller kept tabs on exact timelines —FEMA's Eric Gentry talked to the elected officials and then directly to the man everybody calls Hooty. "Most people don't even know my real name," Adam says. "It's a nickname my grandfather gave me when I was about five years old, and it just stuck."

"Hey! Y'all need to think about evacuating this building," Gentry told Hooty Adam.

"First of all, where would you like me to go?" asked the 42-year-old disaster preparedness director. He took a deep breath and looked directly at his federal counterpart. "And second, I'm not leaving because unless you can tell me I can get my phone lines from this building, I'm not leaving."

"Well, do you have a death wish?" Gentry countered.

"No, I have a *live* wish with three boys and a wife. I don't have a death wish, and I'm not leaving."

But most of the hundred-plus people went.

"They were ordered to leave the EOC, to go to their homes or report to a safe shelter, to Stennis Space Center," Keller said. "And it wasn't five minutes later—I understand; I realize they don't have a choice; they were ordered."

Stennis Space Center—a 13,800-acre test site west and north of the Hancock County EOC. In the early 1960s, Stennis originated as a static test facility for launch vehicles to be used in the Apollo manned lunar landing program. Since, the Hancock County facility has become multidisciplinary, comprised of NASA and more than 40 other resident agencies engaged in space and environmental programs and the national defense, including the US Navy's world-class oceanographic research community. In 2005, Stennis served South Mississippi as a safe zone for pre-deployment in anticipation of Katrina.

"FEMA pulled out," Adam noted. "The National Guard pulled out. MEMA pulled out—every state agency pulled out and left 35 of us in our building."

Chancery Clerk Keller stayed, as did the emergency management staff, some volunteers, firemen, and some policemen. Supervisor David Yarbrough stayed, too, as did an AMR employee.

Keller well remembers Gentry's stance and reasoning.

"His position was, 'This is not the place you'll need to be. I've walked around here, and this building is not made for a hurricane. Y'all need to rethink this, to start your process to move the EOC somewhere other than here.' I was quick to tell him that this building had stood what hurricane legend Wade Guice labeled the Mother of All Hurricanes, Camille, and, better than that, it stood at 27 feet elevation; so regardless of how bad it got, we'd be okay. Where we are is the highest point on the Mississippi Gulf Coast: ground level, slab, 27 feet above sea level.

"He said the projections have this storm coming in at 24 to 26 feet storm surge, and there's going to be 10- to 12-foot tidal waves," Keller admitted. "That puts 34 to 38 feet of water—water into this courthouse and several feet underwater.

"This was presented to Hooty and me," Keller remembers. "We sat with the supervisors and mayors and said, 'Gentry wants us to reconsider, to leave.'"

Representatives of the City of Bay Saint Louis and the City of Waveland quickly broke off to discuss within their own groups. In short order and in accord, they returned and announced, "The county can load up and leave us if you want to because Hancock extends way beyond the city limits, but we're staying."

Gentry tried again to discourage staying. "I want you to do this before you make your final decision," he implored. "Get your laptop and pull up this website about the SLOSH—Sea, Lake, and Overland Surges from Hurricanes (SLOSH)—model."

"What I'm seeing," Keller told others in the war room, "is that the entire south end of the county is going to be under water: all of Pearlington is going to be under water; all of Clermont Harbor, all of Lakeshore, all of Waveland. One little spot at Bay Saint Louis won't be—I don't understand why, but they're showing it's not going to be under water. We're going to be under water all the way to the Interstate (I-10), north of the Interstate all the way to Dolly's Quick Stop at Kiln."

One of the supervisors loudly interrupted, "Bullshit!"

The soft-spoken Keller admitted, "It might be bullshit, but what I'm looking at on the SLOSH model tells me if we don't plan and prepare for it, then shame on us."

Then Gentry asked, "What's your backup location?"

Non-existent, the locals shrugged. Hancock County's emergency management plan had just one location for the disaster preparedness responders, and that's where they had gathered.

"If you won't leave," Gentry urged, "at least start moving your vehicles and equipment north to avoid the storm surge."

Of the 35 souls remaining, most decided to go home that night to get a good night's sleep and return Sunday at 8 am.

"I woke up at 6," Keller would tell his colleagues later. "Watched some TV—the reports that morning had Hurricane Katrina at 125 miles an hour. That was the first time I saw several models showing the western Mississippi Gulf Coast and eastern Louisiana as the hurricane's target. This is about the fifth hurricane since I've been county administrator that we've activated the EOC. All the times before, I've gotten my air mattress, my blankets, my pillow, and enough clothes for four days; no way would we ever need more than that!

"And on the drive in this morning," he continued, "I started taking in all the scenery, realizing that was probably the last time that I'll ever see it looking like that if the projections do happen." His home four miles from Picayune—as the crow flies, about 25 miles from the Coast—provided safe haven for wife and children, his mother, and some friends.

Driving down Highway 603 all the way from Kiln to Highway 90, Keller enjoyed what he remembers as "a beautiful summer day," but he also observed vehicles parked along the side of the road—hundreds of vehicles. He realized that even though evacuation orders went out Saturday night, a lot of people had chosen to stay. Too many who lived in riverside communities, along the canals throughout the lowlands—too many of them moved vehicles to what they thought would be a safe place and then returned to their homes. That troubled him.

On the way, he telephoned the man everybody knows as Boss Hog, a Sheriff's Office employee with a broken foot. "You've never left me before," Keller said upon learning Boss Hog and his wife aimed to head up to North

Carolina. "And I can't believe you'd leave me now—you're the cook. If I've ever needed you, I need you now.

"So he said, 'Well, I'll see you down there,'" Keller recounted. "I said it might not last but a few days, and he showed up. Ninety-three straight days later, he got released from duty."

Adam ran interference, too. One of the AMR employees who lived in Bay Saint Louis became obviously concerned during the evacuation. His wife and family had decided late to leave, and they stalled in traffic. They couldn't get out of town, could not go anywhere because traffic backed up so bad, and returning to their home would have put them directly into harm's way.

Adam knew the problem all too well. "At the time we only had one shelter, and it only held about 200 people and was full. We didn't have a lot of room to put people. My family got out a lot earlier; they were in Tallahassee. So I gave the AMR employee my keys—of course, I knew the family. I grew up with them, and I said, 'The only thing I ask is, don't kill my dog.' I'm actually glad I could help them, and they were very thankful."

Adam also had already helped well-known local ham and low-power radio station WQRZ organizer and operator Brice Phillips.

"Uh, we're not gonna make it," Phillips had telephoned from his house that survived Camille. "It's at the mouth of the Mississippi, and they're reporting 40 to 60-foot wave heights at the NOAA bouys out there. That is not good. It's telling me right there to 'Get out!' So we're coming there."

Phillips got most of his ham gear, all his little radios, knick-knacks, and multiple spools of co-axial cables and then drove to the EOC. Soon he began broadcasting: "We're at the Hancock County emergency management agency right now. We're about 21 to 30 feet above sea level. The military, FEMA, and everybody else is bugging out, but we're going to stay here a little bit longer and see how bad it gets."

Set up in the building's front, alone, where all was quiet, Phillips went to sleep about midnight Sunday and got up at 4 because "that's when they produce the National Weather Service bulletin. I went to the four o'clock advisory to see where it was because"— and interrupting himself that very moment, he growled, "Rrr-run."

"We knew we were ground zero then," he said.

Telephones rang incessantly.

Keller said, "This is not what we need, not what we want. They're asking about sand bags, shelter, and whether they should stay or leave. They're still using Camille as a reference because Wade Guice called it the Mother of All Hurricanes. Even we were ordered to leave here last night. But we would not go—we could not leave. And now other people who didn't go are showing up. They're starting to show up here at the EOC."

Hancock County's one Red Cross-approved shelter at Hancock North Central School was full. "Stennis is sheltering people, but they will not allow us to put that on the air. They cannot declare themselves an evacuation shelter, even though they're not turning anyone away at the gate or at the door."

About midnight, he would recall, two men—one a double amputee—showed up. "Y'all can't stay here," Keller admonished as he convinced an AMR crew to deliver the two to Hancock Medical Center. At two o'clock, he urged everybody to get some rest, if only for a short few hours.

As daylight emerged Monday, the stalwarts who stayed began to see the storm's fury. They braved outside, still under the building's cover but splattered with wind-driven sheets of rain. Boss Hog started cooking bacon, and the telephones kept ringing. Police and firemen returned from checkpoints and reported flooding everywhere; EOC staff quit even trying to list all the streets and just acknowledged that both Bay Saint Louis and Waveland experienced widespread flooding. Television reporters cited 165-mile-an-hour winds and a storm surge of 27 feet.

"We're in trouble," Keller said.

A female EOC worker heard the roof rattling and opened the door on the southeast end of the building. Immediately the wind tore metal hinges off the wall and almost thrust her into the storm. She clung to keep from being sucked outdoors. Others scrambled to catch her, pull her back inside, and secure the door with bungee cords.

Adam was on the telephone with his counterpart in Jackson County, on the eastern side of the Mississippi Gulf Coast. "I'm getting water over here," Hooty heard. "Well, that's funny because I'm not getting any water, and I'm supposed to be ground zero!" The connection snapped, and he walked to the EOC's back door, looking to the left, toward the beach, even though he knew he would not be able to see the beach. Only mildly astonished, he did see water.

"I don't know how deep it is, but it's good enough that it's rushing," Adam told the staff. "We're fixing to get hit with water. We need to get up and move more toward the justice court side, toward the west, because it's a little better elevated than here. You'll drown in here if we get enough water; it's not safe."

As Adam and Keller moved folks frontward, they realized they did not have a life jacket for everyone. Three men volunteered to wade in over waist-deep water to a boat on a trailer out back and retrieve a few jackets.

Phillips was on the radio. "It's coming. We're having 165-mile-an-hour winds. The light poles are still standing, but they're at a 45 degree angle. People are still calling us and asking what to do. I'm sorry. We can't help you. . .

"People are climbing into their attics," Phillips announced, mentioning that one man specifically said he would use an ax to cut his way into the attic. He quoted the caller: "We thank God and hope we can get through this. And if we don't and someone finds us, we—uh—we'll see them in Heaven."

Keller took a phone call from the Waveland jailer informing him that water was coming inside. "Get everybody upstairs to the second floor. What? Houses are floating down the railroad tracks toward Waveland?"

As public information officer, Keller did another live telephone interview with WLOX-TV in Biloxi. He could hear alarms going off as other colleagues tried to shut them off. Then he noticed water pouring down the east wall, getting onto the computers. The building's flat roof had begun to fail. After checking in one more time with his family about 8:30, he took another WLOX call at 9 am that Monday: "The roof is lifting off the EOC," he reported. "What if? What if the worst happens? We'll all go to the safe room."

That was the last anyone outside heard from Hancock County officials.

As the team moved toward the safe room—an add-on designed to withstand 200-mile-an-hour wind—Boss Hog instructed them to leave personal belongings and just get inside. Water was filling the streets, and quick peeks out indicated it was rising at two feet every 12 minutes.

Somebody with a cell phone and video capability captured a fireman: "We're in the safe room at Hancock County, Mississippi. Winds are gusting outside now about 140 to 145, and the eyewall's still a little ways away. We lost all our computers and everything; so we don't have any update. All we can do is look outside and take our best guess. We've got trees bent over sideways, snapping in half, and roofs are getting blown off—the eye is very close."

From inside, they could not see but could hear howling wind and turbulent water lapping against the building, its force tossing untethered objects against whatever was in the path.

Water started coming under the door, Keller noticed. "We don't want to panic everybody," he told Adam, "but we can't stay in here. We'll drown."

"We've got a lot of people here," Adam stretched his arms overhead, locking his hands together and bending to wipe his face with his right arm. "This may sound crazy, but we need to start somewhere and pick a number and go right on down the line. Remember what your number is. This thing could get tricky."

Individually, the 35 people called out, "One." "Two." "Three. . ." and then each signed a roster, wrote the corresponding number on hand or arm, and prayed that they would not need to be identified from reference to the list they secured high on a dry wall.

"We don't know how much water we're going to get," Adam said. "But if something does happen, they'll be able to identify us."

He checked the building and looked outside when he could. With foresight, he and others captured video images as they watched water come up to the front of the building. Every vehicle in the parking lot flooded, but water never entered the facility on that side—only through the back or south-facing doors.

Water also infiltrated the EOC, ankle-deep in the safe room and then waist deep in other areas. Nobody was exactly comfortable in the repurposed bowling alley.

Even from inside the middle of the building, Phillips could hear the rain. He kept broadcasting: "It's a flat-topped building. Concrete block with a flat roof. And the storm's forcing rain into the building and down the walls inside. All the electrical system is mounted on the wall, all the way down the building. And we're going. . ."

He opened his mouth as if trying to bite air, both recalling the pressurization and demonstrating how he dealt with the pain. He pointed to his own ears and yelled, "Everybody in the building. We're—the pressure's unbelievable. Things are not safe anymore!"

Adam also noted the discomfort. "All of a sudden, everybody's ears started popping, and it's like, 'Man!' . . . The eye's passing."

Phillips guessed the time to be about 9 am. Hurricane records would show the double eyewall passed over the entire Mississippi Gulf Coast about 10:30, five hours after Katrina's winds began to hammer the shoreline and inland.

"I wouldn't even begin to be able to guess what time it was or how long it was," Adam would remember. "There ain't no sense of time when you've got water and you don't know if you're going to die or not. You don't know time."

Finally, Keller saw a glimmer of light under the door on the west.

"Just a little flash of light," he pointed. "I realized we had hit the peak, and we had been spared. We're okay now."

Chapter 9

A young family of three who lived a block from the beach had also eyed the approaching storm from Thursday forward, but they remained far from relieved that awful August morning. Just months earlier, on Easter weekend 2005, they moved into their dream-come-true split-level home in West Gulfport. After six months of renovating and adding on to the Frank Lloyd Wright-style house, they took no delight in any potential storm.

Doug Handshoe's family moved to Waveland in 1969, and they survived Camille. As he grew up on the Coast, he and buddies defied the rules to swim in the waters of the Mississippi Sound during many tropical storms and to shoot the "rapids" in the often rain-swollen Wolf River. Such antics almost resulted in their arrest before Hurricane Frederick in 1979. Foolhardy behavior carried him toward Katrina.

Grown up and in business, he did pay attention. Beating the crowds before last-minute shopping started, he with wife, Jennifer, and seven-year-old son, Eric stocked up on hurricane necessities on Saturday—food, water, gasoline for the vehicles. Eyes and ears focused on the weather as they prepared their new home for the expected hurricane. By Sunday evening, most of their neighbors had evacuated so that only the huge old live oaks and magnolia trees facing the Gulf of Mexico kept them company.

Handshoe heard from family and friends from across the country: "Please leave." He assured all they were prepared. A certified public accountant in duo practice, he had conferred with his business partner and her husband, who lived 40 miles inland. His wife, a public health co-worker with the partner's husband, communicated with her colleagues, too. By bedtime, he had the house ready and both vehicles loaded with clothes, tools, and a chain saw—just in case he had to cut their way back in. Simply dozing on the sofa throughout the night, he awakened about 4 am Monday when electric power failed. Without rousing wife and son, he walked onto the carport, noting that a few big limbs had fallen, to be expected with a tropical storm.

Not long after she got up, Handshoe's wife and he saw a large water oak fall from their neighbor's yard to block their driveway. They could not leave. Then one of their own live oaks fell, and they saw their back porch lifting and falling. A wind gust lifted the porch roof completely and flipped it over the back roof. Not one window broke, but the damage had begun.

Awakening their son and telephoning his mother at her evacuation location in Panama City, Florida, the Handshoes learned about 6:30 a.m. that Katrina was definitely headed their way. They remained calm as the weather worsened, and Handshoe suddenly instructed both to "get dressed, put on your shoes, just in case."

Based on his Camille experience, Handshoe expected about a foot of water to enter their home. But looking through the glass door to his back porch, he saw at least two feet of water. Living room furniture began floating, and they hurriedly tried to salvage special possessions. From the master bath, he saw a huge diesel tank floating in the back yard just before it popped through the wall, ramming their new whirlpool tub. They knew then that they definitely could not get away and called to tell their respective mothers; expect another four to six hours of hurricane force winds, they heard from both.

Above the wind's howl, Handshoe hollered, "Okay, guys. We may have to leave!" Opening a window on the north, he touched the glass at the same time a tree limb broke it. He suffered a cut thumb and screamed, but neither wife nor son heard. The fuel tank began to ram the bedroom wall. "We have to get out of here!"

In almost four feet of rising water, the family struggled to save both themselves and their cat. The only viable exit, he considered, was through a broken window in Eric's room on the north side. Pushing through the flooded hallway, they entered a room whose entire north wall was gone and discovered that one of their vehicles had swum from the home's south side to just outside the remains of that north wall.

'Oh, my God!" Jennifer exclaimed. "Look!"

A cargo container from the Port of Gulfport smashed the house across the street to their northwest. As waves swept through the bedroom to the ceiling, young Eric calmly announced, "We are going to die."

Handshoe responded curtly: "No we are not going to die! We are going to keep our wits about us, and we will live. Remember to keep situational awareness. If we don't panic, we will make it through this."

Even he believed the affirmation, and they watched another cargo container move up their street.

Freeing the cat to save herself, the three managed to swim and climb aboard the top of their vehicle just before jumping onto the larger and possibly more secure roof of their house. Eric's foot caught in the debris, eliciting a scream from his mother and fast adrenalin-spurred action by his father to get all of them ready for what they now call "roof surfing."

Amidst tree branches, debris, and another oil tanker, the Handshoes floated north. Then the tidal waves rolled in.

"Big wave!" Handshoe yelled, grabbing both wife and son securely for the 15-foot swell. Amazingly, the roof rode the waves, one after another, each smashing against the nearby floating houses. Losing count after ten big ones, he hugged more tightly and confessed, "I know I haven't always been there like I should have, but know I love you."

Yet another huge wave snapped his mood. "Watch this!" he yelled, struggling to stand against the next swell and yelling as loudly as he possibly could. "Yeeeeee-haaaa, motherfucker! You ain't gonna kill us, you bitch!" Exhausted and with a four-inch gash along his thigh, he dropped back to the rooftop, realizing that his outburst had tickled the youngster whose daddy had said a "bad word." Half smiling, Eric giggled: "Daddy, you're crazy."

Anticipating when the water might recede, Handshoe planned ahead to keep the threesome from getting swept out into the Sound. Before he could take action, though, the tanker truck broke loose and pushed them against a blue house pinned against trees to the north. To avoid getting crushed and battling yet another wave, he managed to lead them over the floating debris pile to the home that had belonged to the Irby couple. Noting but unable to attend to new cuts on his legs, Handshoe kicked in the west-facing window, cut himself again, and pushed away the shards for wife and child to climb through.

All three collapsed onto the bed floating in the room. Bloodied but regaining composure, Handshoe said, "I think we're going to make it," and asked his wife with fear, "What happened to your shoes?" Gone, she said, before they even got out of their own bedroom. Then she opened a closet door, discovering that Mrs. Irby's shoes just fit her and also that some scraps of fabric could become a "battlefield dressing" for her husband's leg and thumb.

Finally protected from the sharp rainfall and with the wind still howling, they continued to hear the house creak from pressure of wind and water. Handshoe passed out on the Irbys' bed. After some time—*10 minutes, more?* —he realized the water was gone from inside the house. Unsure, though, that the walls would hold, he could not relax. Rain continued, and even as he faced destruction in every direction, he announced his intention to "come back here to rebuild." One of the Handshoes' dining chairs appeared near the remains of a prized leather sofa. Their beautiful new computer armoire lay smashed against the window. He also noticed chicken leg quarters everywhere in the 15-foot-tall debris pile.

Despite the wind, he scrambled out the window and up the ladder on the back of the tanker truck. Standing atop the tanker, looking to the south, he saw the water remained high and that not a single house stood between the east-west Findley Avenue and the beach. "We have a beach front lot now," he proclaimed, looking down to see himself standing in a pile of chicken leg quarters. "How the hell did these things get up here?"

Winds began to subside, and as the Handshoes made their way to the top of the debris on their street, they spotted somebody to the northwest. Incredulous that any of them survived, they exchanged "we're going to get out of this" determination, encouraged each other, and agreed any of the evacuees would have been glad to share food and water that had not been destroyed. Declining the offer of a "drunk and obnoxious" would-be host, the Handshoes pressed toward the railroad tracks. "We are walking to life. There is no life left here."

Finally they reached the railroad. Three crossing guards looked at the family but without recognition or even cognition. The Handshoes kept walking, facing a mile or so trek to Memorial Hospital of Gulfport. Several cars passed, but the drivers looked away, not quite knowing what to make of the blood- and diesel-drenched man and his companions. Then a returning neighbor driving amidst the destruction acknowledged them and without hesitation agreed to deliver them to the hospital.

Getting out about a block from the east side of the hospital and walking toward the emergency room, the family reached a triage coordinator. "How are you hurt?" he asked. Handshoe pulled up torn shorts to reveal the gash along his thigh; without a word, the man affixed a tag around his neck and called a nurse. Almost immediately, he was ushered past a badly hurt man begging for morphine and into a cloth-partitioned room. "There isn't room for your family in here," one of the nurses said. "I'll take them to the food court where we have our shelter. They'll be safe there."

Nurses and administrative staff streamed in. "Look," Handshoe said to one with a clipboard. "I pay my bills, but I have nothing but the clothes on my back. Can the hospital work with me until I get my life back together?" Smiling at the obviously tired and depleted patient, she said, "Given the circumstances, we won't be pressing you to pay. Just give me your info, and we'll send you a bill later."

After a licensed practical nurse swabbed his wounds with alcohol-soaked gauze, a nurse came and began conversing. "How did you get that?" she nodded toward the gash on his leg. Finally breaking, he sobbed, "I've lost everything I owned." Hugging him, she whispered into the stranger's ear. "You are not alone; I lost my house, too."

"You're mighty brave to hug a wretch like me, covered in blood and diesel fuel," he said, tugging at his now brown shirt and removing it for further examination. "This shirt used to be yellow. . ."

A physician he knew stapled and derma-bonded the cuts, taking a fraction of the time stitches would have required. On his way out, having learned that the badly hurt man from his way in had gotten some pain medication and settled down, he declined a tetanus shot—"others would need it more" —but accepted antibiotic samples and pain medication. "Trust me," the nurse said. "You'll need them later."

Reunited with wife and son in the food court, he found that Eric had become a "Scotsman, wearing a large Memorial hospital t-shirt and nothing else," and that Jennifer had also gotten a new t-shirt. Settling in for the evening, they swapped "war" stories with 30 or so other refugees and formed a new friendship with a man who offered to drive them the next day to wherever they needed. Tearfully, Handshoe thanked him, and thoughtfully anticipated taking the older man's advice: "After you've had a chance to recover, you need to write down your story. It's one worth hearing, and writing it may help you heal."

The family huddled on the cold concrete floor, trying to sleep but not getting much. Fortified with coffee and breakfast their new friend bought for them the next morning, they drove through downtown and eventually reached their own street. Amidst their goodbyes and gratitude to the new friend, Jennifer's mother arrived to take the family to her home. From there, they were able to also let other family know they were alive.

"I knew when I was on that debris pile that my life had forever changed, though I didn't know exactly how," Handshoe would reflect. "Every day since has been a new chapter as we struggle to understand what happened, and we piece together our lives. A wise man once told me that change is the same as opportunity. We are determined to turn our experience with Katrina into something positive."

Chapter 10

Katrina crushed Mississippi's Gulf Coast that Monday. From well before daylight until her landfall about 10 am, she attacked with knock-down winds plus eight to ten inches of unrelenting rain, an unprecedented storm surge, and widespread deep-inland flooding. But even the National Hurricane Center (NHC) does not know exactly how bad the storm was. Precise measurements of the fury do not exist because many gauges failed and also because the storm destroyed buildings that would have provided still-water marks.

NHC's data do indicate that the storm surge was from 24 to 28 feet for about 20 miles centered roughly on the Bay of Saint Louis along the Mississippi coast. The area encompasses the eastern half of Hancock County and the western half of Harrison County, including Waveland, Bay Saint Louis, Pass Christian, and Long Beach. The highest water mark observed for storm surge was 27.8 feet at Pass Christian, on the immediate Gulf Coast just east of the Bay. Data also indicate that the storm surge was 17 to 22 feet along the eastern half of the Mississippi Coast, roughly from Gulfport to Pascagoula. Apparently, the surge penetrated at least six miles inland in many portions of coastal Mississippi and up to 12 miles inland along bays and rivers. Surging water crossed Interstate 10 in many locations.

The monster storm left most Mississippians without electrical power, without telephone and other communication systems, and—to a large degree—without access to normal transportation routes. She destroyed highways, bridges, and railways and cluttered roads with debris and downed timber. NOAA downgraded this Category 5 hurricane to "only" a Cat 3 at landfall, but the devastation surpassed catastrophe.

People hunkered in Harrison County's EOC could not see or even imagine the storm's destruction. They could hear the howling winds, smashing glass, thunderous rain, and what they guessed were tumbling pieces of metal and heavy plastic. They tried to envision what they would see when they could lift the building's steel door and leave the courthouse. They could not.

Still ensconced within the storm shelter, individuals tried to settle their minds and focus on what to do next. Assessment and response, they knew, but each responder would without second thought dive into doing whatever felt right. They would advance into the aftermath on "autopilot." The commander, the health officer, the coroner—each would react as only he had been prepared and conditioned.

§

After landfall and another few hours of unrelenting wind and rain, EOC Commander Spraggins agreed first responders might exit to start search and rescue. With the surge sucked back into the Sound and winds down to about 60 to 65 miles an hour, they emerged into a totally ransacked reality. Spraggins knew before they even opened the doors to discover Katrina's aftermath that every individual would see and process the storm differently. Each would try in his or her own way to make sense of what had happened during the interminable battering with rain, wind, and surge.

Spraggins exited the EOC with Coast natives and Supervisors Connie Rocko and Larry Benefield. Everything was gone. Beyond their foothold inside the courthouse, they had no notion of where they might be. Walking beyond the parking lot, they saw no landmark. Familiar buildings were gone.

"It looks like somebody just took a bunch of toothpicks and threw them down," Rocko described. "It looks like a bomb exploded, and nothing but rubble remains."

Water had mostly subsided around the courthouse, leaving behind a world they hardly recognized: vehicles in the county parking lot smashed with shattered glass covering the ground, trees stripped of leaves and newly decorated with remnants from the harbor and nearby neighborhood, downed utility lines and leaning poles. They saw, but comprehension eluded all three. They learned that policemen had already found a survivor, a young boy, maybe five years old, at the railroad track.

"He was with his mother and father and grandparents, and they had tried to leave but the car wouldn't crank," recounted a police officer. "The water itself started to take over. The mother and the older brother were trying to take care of each other, and the father had the little boy. I don't know how the waves went, but the little boy was sitting behind part of the firehouse there. Undoubtedly the mother and brother washed to the back side of the apartments, and somebody grabbed them out of the water and took them to a hospital. The grandmother and grandfather are both dead, and the father was lodged inside the building, dead."

Spraggins and his team searched and searched. That same day, walking on the debris, near the back quarter of Water's Edge Apartments—"We had three or four killed there, right by the VA Hospital in Gulfport," Spraggins remembered—they heard a faint voice. "Below me were cars; that's how much the debris had stacked up." Looking down from the debris pile, Spraggins searched for the sound's source. Finally returning to the center part of the pile, he looked up to the third floor and saw an 84-year-old woman standing with her walker.

"Stay right where you are—are you okay?" he yelled. Nearing her, he saw a couple more people. "Just as sweet as she could be, she said, 'Yes, I'm fine. Come on in. Do you need something to eat? Do you need some water?' Let me tell you something: this little lady might not have been in the greatest of health in the world, but she was taken care of. I went in there. She had water, everything—she even had coffee made, and she offered to make me some. But she had no way to talk to anybody. She wasn't worried about herself; all she wanted to do was let her sister know she was okay, and her sister was in Colorado. She had gone to spend some time with her daughter in Colorado, and this little lady wanted to let them know she was okay. So I got her name and number and promised I would let them know.

"We sent somebody back to get her—she was fine, not upset, not in shock."

The officials saw their first looters. "And that griped me," Spraggins recalled. "There were people out there with shopping carts trying to loot, but I really didn't have time to fool with that, and our police officers didn't either. We were taking care of needs."

They tried to go to the beach, thinking they could get there in Spraggins' vehicle. Destroyed, without glass, and full of water, the vehicle was out. Rocko's truck, stashed in the Hancock Bank parking garage a few blocks over, had survived and cranked. "It didn't have a back window, but at least she had known where to hide it. So we finally go to the beach."

Incredulous, they could only stare.

"The water had taken out everything on the coast. We looked to the right and saw the Grand Casino on Highway 90, and the Copa Casino was sitting in the Grand's parking lot. We had known Katrina was big, but we had no idea!"

Soon realizing they could drive nowhere in downtown Gulfport, the threesome abandoned the car and walked toward Second Street. People they encountered "looked worse than deer in headlights. Zombies—that look. It was just a total stare. People were in shock. If you saw it, you'll never forget it. They were trying to talk, and we were trying to help them, but it was just like that. . . We saw hundreds of what looked like zombies, like they were dead but not really. They looked like they could not fathom what had happened to them. What am I going to do?! They had nothing. And they were helpless. They didn't know what to do, and we couldn't do anything for them—if they were walking, not hurting, not bleeding, then we just went on to the next one."

§

Deliberate in thought, word, and action when under pressure, County Health Officer Dr. Robert Travnicek struggled to associate with the new reality. Having been on the telephone most of the night with personnel from area hospitals, unidentifiable voices pleading for help, and also occasionally with his wife from her job at Garden Park Hospital, he needed air. Outdoors for the first time about 4 pm that Monday, he looked and realized "we'd just been through the biggest thing in world history. And it was mainly—these things are always mainly water. We're dealing with a myriad of problems."

His mind raced. No power—hospitals would have to operate on backup generators; even then, nobody could know that fuel would become a resulting emergency issue. Hundreds of dialysis patients would need treatment. What about the wounded? No potable water existed; the sewers ceased functioning because all the lift stations were destroyed, "and I'm sitting next to these people trying to make decisions about what we do, and that's when we are out of. . . You know, basically then, there's this Emergency Support Function 8—and we're responsible."

What he wanted most then was two historians and a photographer. "That's the first thing I have to have. This is not—now we have to think outside the box. This is the biggest thing in history; we can learn from this immediately. We need to have . . . You have to have people with historical perspective, that kind of writing ability, that kind of sort of thing. Send me historians. Send me three public health physicians. I don't want any jerks around here that don't have a medical degree, and it would be nice if they practice medicine, but that's not necessary. . . but they have to—I have to have them deputized; I have to be able to deputize them to condemn bodies, to condemn food.

"The first day, the first five or six days, I think I condemned ten million pounds of seafood," his stream of consciousness continued. "I never even saw it. Normally I like to go see things, but they came to me from EPA or DEQ and there was nobody. . . there was nobody here to help, and so I had to do it myself. People would hand me their handwritten or typewritten notes, and I would condemn five million pounds of shrimp at a certain place—you could smell the shrimp. And so another thing I did, again, I worked with the sewer people, that was out of our hands. I tried to get public health, US Public Health Service, they had a bunch of engineers running around, I don't know what the hell they were doing, they were absolutely worthless, and I tried to get them to, they said that they had people and engineers that were capable of doing equipment lists for the lift stations, that knew what was in these lift stations and get equipment lists so we could get FEMA to come down and get these lift stations going. Trust me, I mean this thing was so complicated, and I'm, you know . . . I wrote a mission statement for US Public Health Service to come and repair the sewers, and Jim Craig took it to the trash—said nothing was going to happen without him."

Travnicek walked outdoors late that day, walked into bright sunlight and a woman he'd known for years as a Red Cross volunteer—a woman desperate to find her daughter Jennifer and her family. The daughter worked for the health department so maybe—maybe, her mother thought—maybe the health officer would have heard from her. She and her husband Doug, a CPA, decided to show their child the storm, the mother described, so they stayed in their newly refurbished home near the beach; he knew from having survived Camille that their home likely would take on no more than a foot of water, she said. They lived solidly in evacuation Zone A, Travnicek realized. "We thought they were dead. We wrote them off . . ."

§

Coroner Gary Hargrove waited inside the bunker until winds dropped from some 130 miles an hour to about 65. He knew it was going to be bad, but he also realized his responsibility to start search and rescue. Grabbing his gear and recruiting Misty Velasquez from telephone duty, the veteran coroner collected body bags from the funeral home.

"Our first job is to rescue those that have called in," he shouted to the young woman who'd answered frantic phone calls throughout the night and halfway through that Monday, until telephone lines died. "We'll go out and do what we can, rescue those we can, and then take the death calls."

On Kelly Avenue, Hargrove and Velasquez ran across a woman walking up in a pair of shorts—muddy, cut, bruised. They got help for her and started walking along Second Street into the neighborhood, walking through what was left, trying to find addresses that had been called in. "Surreal," Velasquez described scenes from both street-level and atop the debris piles. "It was not like anything anybody can explain to you. Full houses backed up and stacked up looking like doll houses, two or three with some of the walls torn off, but the beds are still made, and the rooms are showing." They saw clothing shreds in the trees, cars upside down and piled on top of dismantled structures, concrete slabs where homes once stood— utter catastrophe. They could smell gas leaks and hear people screaming for somebody to scream back and save them from under the debris—"debris taller than your head. And you could feel the nails in the rooftops, but after a while, you don't, because it's just so overwhelming.

"We've got to get to the addresses to try to find and rescue people, but the looters are already out there, too. Gary's truck was in no condition—the side window was out, the back window was out, and it had water inside—

but we just threw the body bags on the seat, on top of the glass and in the rain. . ."

Later, Velasquez recalled, "I don't remember being dry for 48 hours. But I do remember finding a police officer—a police officer crying. We hadn't seen everything but could just see some signs. Gary said, 'Keep going. Don't try to look at it all; it's too much. Just look at one thing at a time.' That's what he told me. Then we had a call for our first recovery.

"We found this little boy—I want to say three to five years old—we found him on the wood piling north of the Veterans' Home by the Security Guard, and he was alive. He said, 'Mamma and brother were swimming with me, but they swept downwards.' We started looking around because we knew we had a mother and son—and there's this empty shed. I was looking and thinking, 'I'm not built for this.' I walked slow because I really, really did not want to find something. Then, sure enough, I saw something, and I thought 'Maybe it's a dog,' but I kept walking and, sure enough, it was a young man, the father of this little boy. He was gone. He was clutching . . . That was my first find. And I called Gary—'Gary, can you come here for a second?' I can't tell you how many times we went back out there, even with dogs, looking for that mother and son. And we kept looking and looking but could not for the life of us find them—and it drove us *nuts*. We were all just haunted by the story of that little boy."

§

Rupert Lacy came out from the bunker to what he expected would have been his "fully protected" service vehicle. Parked between two military vehicles before the storm's landfall, the truck sported a satellite dish from across the street sticking out its back window, a windshield shattered with what could have been buckshot, a drowned digital camera on the front seat, and an expensive but now useless flood coat. The plastic garbage bags he had affixed to protect radio equipment were tattered or gone.

"Secure your stuff—it just wasn't meant to be," the hurricane veteran lamented.

First responders braved that Monday afternoon's continuing wind and rain and then bright sunshine to discover wind-stripped naked people who had hung for hours onto tree limbs for life and battered souls still on the rooftops or straggling through the community; boats atop debris piles; and boats, cars, and trucks heaped like broken toys. They also discovered 30 dead bodies at a beach-side apartment complex in Biloxi. Associated Press reported a spokesman for Harrison County EOC confirmed 50 deaths Monday evening.

Chapter 11

Katrina's eye had passed over Bay Saint Louis and Waveland, and the wind and rain calmed some. But for the emergency management team inside what remained of the Hancock County operations center, the storm would not end.

"People started trying to get out," EOC Commander Brian Adam explained. "They were trying to get to family members. I saw this one truck and a guy whose family I knew growing up. He waded up to our building, and I asked, 'What are you doing?!' And he said, 'I'm trying to get to my sister and them, and I'm going. . .' 'Well, Bud, look how *deep* the water is. You're driving a big Ford but you can't get through it!'"

Not long after they sighted a Suburban over-filled with both adults and "a bunch of kids. They're scared—we've got to go get them. They're trying to get to safety and don't realize they're running into more danger and putting everybody else there, too."

"What's going on?" Brice Phillips exclaimed. "Are they ducks? Come on over here!"

Phillips recalled, "Then we noticed the surge is starting to run out. I'm not talking about slowly receding; it's getting *out* of here. Just like that. It *flushed*."

Then he saw his old van that had brought him and his equipment to the EOC.

"There ain't no roof on my van," he yelled. The storm had "sucked it right off and tossed it onto the ground. Sucked that roof right off that van. Un-be-lievable. . . . Now how am I going to get my ass back online? Getting the broadcast station back online's now my number one priority."

He tried to make sense amidst the chaos: "I'm going in and out the back trying to get the radios up, to bring all the public service in. . . starting now to finally get Bay Saint Louis straggling in. I'm assessing what to do next, what I do for communications and what we have left to work with. And we have nothing."

Storm analysts would try to determine exactly what happened that Monday. University researchers from South Alabama to Wisconsin would pore over Katrina's data, trying to make sense of the catastrophic cyclone. Meteorologist Bill Williams, director of the Coastal Weather Research Center, University of South Alabama, would publish his findings in 2009.

"Katrina was one of those rare hurricanes in which everything came together: the shape of the coast, funneling of the storm surge, the extreme power of her winds, and, most importantly, a double eyewall structure. The double eyewall structure made her a monster storm, and collapsing cores brought extremely high winds down to the surface quickly. Katrina produced massive damage before the water came ashore."

The wind, and then the water. Who could have known?

"I was not prepared for what I saw immediately after," Hancock Countian Tim Keller said of his Katrina experience. "The worst case that I was thinking the day before the storm was, maybe some of these trees will be missing. There was no way to be prepared for what we saw after the thing. It was overwhelming. Over-whelming."

After the eye passed and the storm came back from the other direction—not as strong but still producing some wind and rain—that's when people started getting outside to see their vehicles.

"All the vehicles flooded," Keller commented. "Most of the trunks are up, which indicates they went under water far enough to trigger the safety system so the doors came unlocked and the trunks came up. Windshields are knocked out, but my wife's Suburban—it's okay. No broken window, and it's completely dry. What in the world?"

None of the 35 who endured the storm from inside the EOC had time to fully reflect. People who had stayed in the community took their first opportunity to move, to find other survivors; some drove, many waded, and more boated to the emergency management location on Highway 90.

"A lot of these people were in shock," Keller recalled. "We gave them our dry clothes. . . some who had clung to a flooded building all night and through the storm came in, and some of the women were really shaken, and shaking. They were *cold*. Here it was the very end of August, and they were cold. A lot had to do with shock. I took some of them in and pulled out my clothes and gave all my clothes other than what I had on. They were soaking wet; so I gave it all away. I had no idea it would be five days before I could go home."

Radio Station WQRZ Operator Phillips waited on the wind to work on his tower. "I've got to get up that tower. Hooty's telling me, 'You can't,' but I've *got* to get up that tower."

"Brice, you can't go up that tower," Adam shouted. "The next thing you know, I look around; he's gone. So I go out there and I see him up there—I'm looking way up that tower, and I said, 'What the hell are you doing?!' And he said, 'I gotta fix it; I've got to straighten my antenna.'

"Brice is a *genius* with electronics. And he was running all this on car batteries," the emergency response manager offered. "And he's skinny as a rail—he's not like me; he's not meaty. And I said, 'Yeah, but I can see your obituary now: I survived Katrina, but my dumb ass fell off the antenna in a 50-mile-an-hour wind!'"

For two-and-a-half hours, Phillips struggled to climb 45 to 50 feet above ground and mend the tower so he could resume broadcasting. Adam understood: "He got it fixed and provided the only communications we had with the outside world. He's still a radio man, one with passion, and a very firm supporter of this EOC."

"Now I see all the damage," Phillips would report. "I could get a real bird's eye view of what was going on, and I could actually see it. By the time I came down, I went inside and told Hooty—I let him know: This is serious."

Adam had done his own assessments from ground level, discovering that only two vehicles remained operational. "Mine and one other—it belonged to a guy who worked with us, a big four-wheel drive. They went down to the Bay Saint Louis bridge, four or five blocks from the office, and came back with 'Hey! The Bay Bridge is gone.' I said there's no way in hell that Bay Bridge is gone. They said, 'You need to go look at it.' So I went down there and saw the Bay Bridge completely destroyed. I looked to the right—

and to the right of the Bay Bridge is what we call a bluff. It's anywhere from a 15- to 25-foot bluff, and it starts and goes back toward Waveland, and it stops there about Washington Street. When I looked at those houses along that bluff, most of them were knocked off their foundations. Well, you know what—if this is this bad, there's not a house standing on a flat beach.

"But we can't get down there," he admitted, looking to fellow EOC survivors. "Water's still high, but some people are getting out. We've got to try to make contact with the police and fire departments. We've had no contact at all. . . Our building wasn't knocked down, but everything we had in there is wet. We don't have a telephone or a computer that works. No communication. Line of sight's all we've got. Not even a radio right now; nothing."

One asset they did have, however, had come before the storm and stayed through the horror.

"Two guys from Iberia Parish, Louisiana," Keller remembered. "I couldn't imagine why they were over here, but they came and brought an airboat they secured through the thing."

With two trucks, a commandeered school bus, and the airboat, Hancock County began search and rescue. Keller climbed aboard the airboat and, with the father and son team, began retrieving stragglers from the water.

"Four or five at a time," he recalled. "We would pick people up and take them to Highway 90 and 603; that's where we could get them onto buses and take them to Bay Saint Louis School—not the shelter, just a place we knew we could gather them all. It was affected, but we knew it still stood."

Adam collected the two mayors and, with Supervisor Yarbrough, used the few remaining dry maps to plan search and rescue.

Keller was eager to check along Highway 603, where he'd noted stashed vehicles before the storm. "It was just the strangest thing. Every vehicle at Derussy Motors was floating straight up with their tail lights sticking out of the water. At that point, the water was still at the top of the building, 12 or 14 feet deep. Then I realized how much water we'd taken on.

"I estimate 200 vehicles," he said before seeing nearby bushes move and directing the boat in that direction. He recognized one of several people in a small boat. "We got as many as we could into the airboat, a man with his pajamas on and bare-chested. He recognized me, and we started crying and shaking—he was in shock. He hugged me and cried like a baby."

Delivering those four to the drop-off point, Keller looked to the west, where an underwater building with just the peak of its roof jutting above the water revealed a survivor.

"There was a black lady, barefooted, with short pants and a t-shirt on," he described. "She was jerking and swinging her arms. We pulled up to her and she just jumped over into my arms. She was no doubt in shock; so we took her back to the drop-off, still stomping her feet, clinging, and hanging. It was surreal.

"You've got to separate yourself from the emotions," he reminded himself. "It was just back and forth, back and forth. And finally, we pulled into Longfellow Drive, the water still about six feet deep. There was a big oak tree, and we pulled through so we could see, and we saw three people in what was left of the right of way, climbing over trees, and they had a big ice chest—they were hanging onto each handle, and a guy was on top of the ice chest. We cut the engine off because the airboat engine was so loud, and the wind was still howling. And I heard some metal in the distance. It was ping! Ping! Ping!

"Then I realized it was people over at those apartments—people at Waverly Apartments making the sound."

Getting to the source took some time and trying. Again, Keller asked to kill the engine. "Now I know what it is," he announced. "We've been boating over the tops of vehicles left in the parking lot, with their horns sounding and lights flashing. We've pulled people out, but I'll bet there's more."

They found nine people, people who got stranded in the apartments, people whose cries for help could not be heard over the wind and the engine's noise but whose presence the pings revealed. Those pings came from one of the women's banging together a skillet and some pans she found in the kitchen. "Pretty smart of her," Keller said. "That led us to find those who were scattered around the complex but made it to the second floor. Then we went back to Longfellow Drive and rescued a lady and a gentleman, about middle age. When we got them into the boat, he stopped and pulled off an artificial leg and poured the water out of it. I've gone to see him since—he wanted me to come see him, and I did. They'd been in the water for several

and raised in a community and the school you went to, the church where you got married, everything—you know, a tornado strikes a neighborhood and the neighborhood's gone. But Katrina took *everything*. Every family member I have lost everything, with the exception of my father-in-law; the whole family lost everything. We had one house left."

Realizing they had lost all possessions would come later. Fayard's job that hot, wet Monday was to save lives. To get beyond the "useless," he and an assistant police chief who had stayed at the EOC commandeered a school bus used before the storm to evacuate special needs patients. Only that school bus and one public works vehicle, a backhoe, had not flooded.

"Everything in the city was lost," Fayard said. "My vehicle, all the vehicles parked at the EOC—all underwater. So we got the school bus and drove to the police station."

With up to two feet of water still covering the highway in many places, they drove along the grassy median between the east-bound and west-bound lanes of four-lane Highway 90. They drove over downed power poles and other unknown dangers.

"You see it's bad, but you really can't come to grips with it," Fayard said. "The entire cities of Bay Saint Louis and Waveland, the entire Gulf Coast was wiped out."

Then they learned what happened at the police station, where some 27 officers escaped through the ceiling onto the top of the building. Strong winds swept some of them off the rooftop, and they grabbed onto whatever they could feel—*were those bushes crepe myrtles?*—to save themselves and each other.

"When we got there, we realized they were not police officers," Fayard recalled. "They were victims. Everything was gone. Our city is gone. They all could have perished, and none of them knew anything about their family members."

Along the way, they had picked up people by the highway, people who braved the aftermath to seek something, anything. They loaded the bus, rounded the corner to discover Hancock Medical Center employees standing outside, and learned the emergency room had taken on four-and-a-half feet of water. "No, no!" the workers yelled. "You can't bring them here! The hospital's destroyed!"

"Well, what do you want us to do with them—take them back out on the highway and dump them?"

The hospital victims had no choice but to become first responders. Emergency medical services veteran Fayard listed their resources to motivate their overcoming initial, paralyzing shock. As officers of the community, they had to respond, he urged. "We put together a team with doctors, nurses, respiratory therapists along with myself and Dave, a paramedic, and we worked until about one o'clock that morning up and down Highway 90. Finally, as the water subsided, we were able to get a little way up 603, not far, picking up people who had migrated to the highway and bringing them back to the hospital."

They realized the hospital could take no more, not unless the individual really needed medical attention. They picked Bay High, which did not take on water, as the alternate shelter, put an EMT in charge there, and continued to rescue victims from the highway until nobody else appeared.

Water still surrounded Hancock County, but they managed to get out via Highway 603 to Interstate 10 and return to Gulfport, to Harrison County EOC. They told their stories, recounted the mandatory evacuation for Hancock County, and second-guessed their own decisions. "Why didn't we leave? FEMA told us to leave," Fayard admitted. "Well, in the first place, FEMA has never been right in the history of FEMA! But FEMA hit it on the mark this time. FEMA said, 'There's going to be a 40-foot surge,' and you can look at those pine trees and know we got every bit of a 40-foot surge."

Those who decided to stay knew their area. They knew what a strong southeast wind would do to the water. They knew that "no matter how big something is, if you put enough water under it and wind behind it, it's going to move." They compared current information and historical data. They knew that Gulf water had never before covered the railroad tracks. So they stayed.

"I could not abandon the community," Fayard said. "The leaders of the community are going to leave? I could not. Our leadership made the decision together, and I'm glad I stayed. I would not have jeopardized my people or our equipment, but I could not abandon the community. I think we saved a lot of lives because we did stay."

Chapter 13

From their homes in Vicksburg, National Guardsmen Lieutenant Colonel Rick Martin and Captain John Elfer deployed Sunday afternoon to South Mississippi. Iraq War veterans from one of Mississippi's first units sent over in 2003-2004, they had already been on the Coast in 2005 for Hurricane Dennis, which sputtered out as a tropical storm. As they drove south on Highway 49—no one else southbound except their small caravan of six vehicles—they noted bumper-to-bumper traffic on the northbound lanes, people they assumed to be evacuating the Mississippi Gulf Coast and New Orleans. Their company split on the coast, giving Jackson, Hancock, and Harrison County emergency operations centers their expertise as "military liaison officers."

In the Harrison County EOC at Gulfport, Martin and Elfer measured the mood of fellow first responders. "The storm got kicked up early Monday morning. For the locals, and especially the people who live here and had been through Camille, you could tell there was obviously some worry. I remember on up into the morning, when it got lighter and the water started coming into the building—*man!* We thought it was raining. And Gary Hargrove looked at me and—I never will forget this—he said, 'Man, that is not rain water; that's the Gulf of Mexico.'"

At 10 am, they and others huddled in the EOC's nerve center briefed via telephone with Mississippi Emergency Management Agency in Jackson. "They talked about massive damage in Harrison County and that several more hours of hurricane-force winds would continue."

Sometime later that morning, the phone "went dead" during a call to colleagues in Jackson. They had no luck with satellite phones, either. "We were pretty much cut off. At 1300, at one o'clock, the wind gauge pegged out—broke; the winds were over a hundred miles per hour. And then at three o'clock that afternoon, the winds were down to 60. That's when John and I went out to assess damage via patrol." With many roads still under water, they could not get to the National Guard Armory north of Pass Road.

Back inside the EOC, they received calls for help. "At four, Pass Christian requested search and rescue missions, and the same thing from d'Iberville," Martin reviewed from his log. "At five o'clock, we had a request from Gulfport Police Department for MP support to help with looters—there was supposed to be mass looting in Gulfport—and Biloxi requested military police support for looters." And they received a request to help secure the hospital.

"For some reason," Elfer interjected, "a lot of the population seemed to migrate toward the hospital, which became a crowd control problem and a security issue for the hospital. Gulfport PD and Harrison SO were doing everything they could possibly do."

Martin continued: "We had requests there from all the different communities for assistance, and we sent everything out. At seven o'clock, John and I went back out and checked on the MPs at the hospital on Pass Road, and at 9:30 I took General Spraggins and some media personnel to the Pass School shelter on Pass Road. At 11 o'clock we returned to the EOC, and General Cross had a request at 11—a mission request for MPs. There were multiple requests—I've got Harrison Sheriff's Office, Gulfport PD, Long Beach, Biloxi. . . some of that is search and rescue, the majority of it. But there were some Gulfport and Biloxi requests for support for looter bands, 25 to 30."

Elfer described walking outside that night. "I remember how dark it was. It was as dark as it was in the desert—no light. None. Zero. Nothing. Just people—almost like a movie—just absolutely in shock from what had happened."

"At four o'clock in the morning on the 30th of August, the MDOT (Mississippi Department of Transportation) commissioner and staff arrived," Major Martin recounted, "and also the main National Guard force showed up. It took seven hours to cut through trees and such from Hattiesburg to

Wiggins—there was so much damage. They were trying to reach us the whole night and got here that morning at four o'clock. They went through two backhoes trying to get everything out of the way."

The Harrison County assignees helped national media reach storm survivors who "had weathered the thing on their roof," Elfer said. "Most of the media were set up at Highways 49 and 90, where all of that debris and the Dole refrigerator trucks—all that stuff that was in the harbor, all that got washed up to the railroad tracks, the casinos and all that. It was a maze to get through that. But the 890th engineers—I've got to give them credit. They had a lot to do with getting Highway 90 back open. They blazed a trail from Biloxi to Pass Christian.

Martin explained: "Lieutenant Colonel Johnny Sellers was working very closely with General Spraggins. He had a plan of attack. You have to remember, those people are from that area, and a lot of them had had their own homes damaged. They hadn't been long back from Iraq, and they left their homes and came to support the hurricane—that says a lot about their character, I think."

Elfer continued: "General Spraggins—since he had been prior service and he was friends and both professional and personal associates with people like Colonel Sellers—he understood what the Army had, what the military had. And I don't want to leave out the Seabees because we had a really good liaison that was absolutely one of the best people and one of the most professional military officers I've been around. The Seabees were absolutely wonderful."

They learned later that the Seabees, the Navy Construction Batallion, had a plan, if needed, to rescue the responders in the Gulfport EOC. Some personnel who would have been involved in that "actually saved a number of people from a townhouse in Back Bay Biloxi."

"We say that to say this," Martin said. "General Spraggins and those people who had been in the military knew what resources were available and knew how we operated as far as command and control, and we really were their go-to people."

With no communication system, they relied on hand-delivered requests for help. "This is down in the ditches," Martin admitted. "911 calls and then you to me, 'Hey!—This is what we need. Do you have?' We were way out there on this. This is not how it's supposed to work. The request is supposed to go in to MEMA or FEMA, and they validate the request and then they give it to us as military liaison officers, and we turn around and tell people at the forward operating center that this is validated. . . Go forth and do good things. But we were on the short list. And that was going on for a little while."

Chapter 14

Mississippi tourism efforts tout "62 miles of scenic shoreline that includes 26 miles of soft white sand . . . gentle waves that dance on the water's surface offering an exceptional view from US Highway 90, aka Beach Boulevard."

When Katrina landed along that coastline in 2005, Bobby Weaver was responsible for protection and upkeep. As director of Harrison County's Sand Beach Authority, he worked pre-storm to prepare the beach, to remove all vendors, signs, and chairs and umbrellas, paddleboats, and other recreational amenities. Then he went home, confident that all would be well and that the day after the hurricane, he and his crews would work to identify damages to roads and bridges, document the destruction to FEMA, and restore Mississippi's prime tourist attraction.

Not until mid-morning Tuesday, the day after the storm's landfall, could he get from his home north of I-10 in d'Iberville to the courthouse in Gulfport. Debris and downed trees littered the landscape and made movement via normal traffic routes all but impossible. Slowly, Weaver drove west, daunted by the total devastation he saw in every direction. Arriving amidst the destruction and disarray at the county seat, he learned from Supervisor Larry Benefield that colleagues had picked him to be their operations chief.

"Okay," he said with what he knew to be a "deer in headlights" look on his face.

Working the EOC for the first time during a disaster event, he sat with Deputy Sheriff Rupert Lacy and Mike Beeman from FEMA. With no option except on-the-job training, he plunged with them into talking about where the county could locate points of distribution (PODs) for emergency supplies they expected to arrive for residents who had survived the storm. The arrival of "two gentlemen – a taller gentleman and a shorter guy" interrupted the session. Representatives of the responding incident management team from Jacksonville, Florida, had arrived.

"They looked familiar so we just stopped our meeting and looked at them and asked if we could help them," Weaver remembered. "And they said, 'Yeah. We were sent over by Governor Bush to help y'all set up the emergency operations center because the report was that Harrison County was destroyed and everybody else was gone.' 'Well,' we said, 'Obviously we're not destroyed, and we're here; but we'll be glad to accept your help.' When I look back on it, the first two or three days—that team's being here was extremely instrumental in our addressing the response, simply because they were not emotionally attached to the disaster. And we were."

Mississippi Emergency Management Agency (MEMA) from its headquarters in Jackson had been communicating with Florida Emergency Management Agency in Tallahassee, standard operating procedure under the nation's Emergency Management Assistance Compact (EMAC). Part of the National Response Plan, EMAC complements the federal disaster response system, assuring timely and "seamless" movement of both supplies and personnel to a disaster site.

The newcomers deployed from Jacksonville, Florida, before Katrina's landfall in Mississippi. Florida and Texas had "stood up" before the storm to fortify the region. Using the Mississippi River as dividing line, Texas would handle Louisiana west and Florida would handle Alabama and Mississippi. The Florida team moved along I-10 as a "perfect work corridor." They got fuel in Pensacola and picked up as part of their convoy some public works people from Harrison County, Mississippi, who had evacuated before the storm.

Moving through Alabama and getting into Mississippi revealed catastrophic destruction. Littered bridges and some stretches of highway still under two or three feet of water slowed their travel. They drove around sailboats on the interstate and learned later that safety engineers closed sections they traveled but then determined to have been too damaged for travel.

Sensitive to the plight of the people they had come to help, the northeast Florida team pulled off I-10 onto Highway 49 and stopped to re-group. They intended to go in with respect for the locals' situation and also as a team; they donned identical t-shirts that proclaimed their role as emergency responders and sent two ahead to make initial contact with the local authorities.

Incident Management Team Commander Chip Patterson and his deputy Martin Senterfitt met General Spraggins first. "We told him that we were there to do anything we could to help, even taking out the trash. We emphasized from the top: We are here to assist your county for your county to be successful.

"We met Mike Beeman," Senterfitt said, "and he explained that General Spraggins was doing an incredible job dealing with the storm impact. And we were told that day that only about 26 to 28 percent of county workers were still in the county and that they were extremely short-handed. They introduced us to Bobby and Rupert, and the first thing we did was to explain our job was to support them."

Patterson as team leader connected with Spraggins. "As plans chief for my team, my job was to make sure things were orderly and organized," Senterfitt said. "So I asked, who's your plans chief? Your ops chief? Your logistics chief? 'Well, we don't really have that,' Rupert admitted, and he said, 'Look: y'all are pretty much going to have to help us build from the ground up. A lot of these people haven't had incident command training.'"

With only brief hesitation, Senterfitt asked, "Who's the best person to get stuff—who knows everybody and can get stuff? They told us the person who knows people and can get stuff moving was Bobby Weaver, and he was already doing a great job at that, so we made him the operations section chief. The one place they didn't have anybody was in planning; so they moved me to plans chief for the whole event, which put it on my shoulders to make sure everything was organized, documented, and tracked appropriately. That's how the initial command structure started, and we got organized. We got everybody in these right positions so we could start moving forward."

Mike Beeman documented the decisions as he drew the structure on the everyone.' So we said, 'No problem; we hereby create a fueling unit,' and we put a guy from Jacksonville in charge of fuel. The guy said, 'Well I don't really know anything about fuel or this county—he's a fire department chief—so I said, 'Well, learn everything there is because you're in charge.' So he took over dealing with anything having to do with fuel so Rupert could deal with other problems.

"As Bobby ran into operational problems, we would ask Bobby, 'What's your biggest problem?' He would say what his biggest problem was, and we would find a person to take over that problem, and that person would go off and deal with that problem. . . We saw a problem with getting aid out to the people—we took one of the people from our team, and he went forward to help with anything dealing with food and feeding. Every time we saw we were dealing with an overwhelming problem, we would get on the phone via the satellite system back to Florida and say, 'Send us 15 more people, send us 20 more people.' On Day One after the storm, on Tuesday, our command structure had about six people; by day nine, we had 65 people. We kept expanding the command structure to take on the problems we were dealing with and making sure that everything would happen."

Among important tools the Florida team brought with them was a refurbished US Navy 45-foot semi-trailer communications van which, Beeman said, "became critical to the response operations because all hard line and cell phone communications capability in the county was lost" for more than a month after the storm. The self-sustaining van provided a large command/conference area, a four-station communications area, satellite TV, satellite Internet system, satellite radio/telephone system, and a 20 kilowatt generator.

When the Floridians realized General Spraggins' increasing frustration about supplies and people not getting to Gulfport in response to the storm's destruction, they introduced him to the communications capacity within the van.

"We took him into the tractor-trailer command post, and we sat him in front of CNN or FOX news—of course, no one in Mississippi at that time had a TV," Senterfitt said. "He sat down for about 20 minutes and watched live coverage of what was going on in Louisiana, the failure of the water systems in Louisiana, the looting, the rioting, the fires. I remember seeing tears in his eyes and he said, 'Well I understand now.' And he went back into the EOC and called everybody together and said, 'Let me tell you what's going on in New Orleans right now. Let me tell you what's happening. It's worse there than here, and that's why we're not getting a lot of aid. It's all going to try to save New Orleans and Louisiana.'

"One of the proudest things I saw was when he looked at the group and he said, 'Look folks: We're Mississippi. We know how to stand on our own two feet. New Orleans needs it more than we do right now. We can do this ourselves.' The whole room—instead of people in Harrison County getting angry or upset, there was a sense of personal responsibility and strength and resolve that passed through the room. Everybody just looked at each other and said, 'Well, okay. We can do this.' It was one of those classic moments in history when you're just so proud to be an American—it was a can-do attitude. There was no complaining, no whining. It was, 'We're going to step up and do this ourselves.' And that's exactly what happened."

By late afternoon Monday, people began to show up at the hospital. Tammy Boudreaux was in charge of patient registration. "What is your name and address?" she would ask. "I don't have an address. I don't have a next of kin. My mother just died in the storm."

Irene Brown, RN, director of women and children services, recalled a child discovered near the VA at Gulfport and brought to the hospital. Stephanie Gable, RN, told the story: "The manager of the mother-baby floor came by the desk pushing a little boy in a car that we have in pediatrics. And she said, 'Stephanie, I've brought you a little boy.' She said, 'His name's Justin, and he was found outside there at the VA in some rubble. We're not sure where his family is. . . Would you mind taking care of him for a little while?' And I said, 'Absolutely not.' Justin and I colored all evening, and he told me his mother, father, grandmother, grandfather, brother, and dog Andy were in the apartment and that they were swimming to Georgia. We weren't sure what that meant, but we had an idea. Later on in the evening, he said, 'You know, I don't think my mommy and daddy are coming back from Georgia.' I asked him why, and he said, 'Because they drowned.' 'Tell me how—why you think that.' And he said, 'Because my mommy can't swim, and I saw her go under.' So I kept coloring and tried really hard not to cry. As the evening progressed, he told us a little more of the story. The next morning, he woke up, and the first thing he did was smile. And I thought, if he could lose his family and wake up in a strange place with a stranger lying beside him—then it was going to be okay. No matter what my house looked like, or if I even had a house. If Justin could smile, I could, too. I remember vividly and was amazed at that.

"That afternoon, we got a call from the emergency room," she continued the story. "Judy was at the desk and said, 'What?!' And she grabbed my arm and said, 'They think they found Justin's mother. She's in the emergency room.' So we went down. We were afraid maybe somebody had heard it on the radio, and maybe it was an imposter. We felt like we had talked to Justin enough to tell whether this woman was telling the truth or not, and—there was a little boy who was an older version of Justin. Judy asked, 'Are you Cody?' and he said, 'Yes, M'am.' That was Justin's brother and his mother! They had floated. They all went under as Justin had said. And when they came up, she grabbed hold of a BFI garbage can, and they held on to that for about six hours until the water receded. And rescuers found them. They were in a shelter that night talking about—I'm not sure if she was crying or telling them she had lost her son—and someone had heard the radio announcement and said, 'Your son's at Memorial.'"

From others who appeared, the staff learned how people had clung to rooftops, suffered cuts and broken arms and legs, and been washed out of their houses. One elderly woman rescued off her roof had been bed-ridden and stayed in her home. The water rose, floated her in her bed out the front door, and put her onto her own roof, still in her own bed. After three days, a helicopter rescued her, her rosary in one hand and a bottle of Southern Comfort in the other.

Sometimes four helicopters, one after the other or concurrently, would deposit people. "Little old ladies curled up in the fetal position in rescue baskets," a nurse remembered. "We would catch the baskets, put them on the gurney, and get them into the ER. The helicopters would not cut their rotors but would just deposit the patients and leave."

Another mother and her son were flown in from Clermont Harbor in Hancock County. "She was like 86 years old and had been hanging in a tree for at least 10 to 12 hours," her rescuer said. "She had bruises from the top of her head to the bottom of her feet, and we cleaned the mud off both of them—they were just covered in mud. She kept talking about her feet touching the clothesline; she was hanging onto some piece of debris. The ambulance guy told us it wasn't a clothesline but the telephone and power lines—she was up that high!"

One man arrived covered in bark—his chest, arms, and legs, all covered. He had held on to a tree to survive. Another was really sick, conscious, but had been exposed for a day or a day-and-a-half. "He asked about his family," a nurse said, "and another nurse behind him signaled that his wife had not survived. He didn't know—and that was probably the most human moment I had. He didn't make it." A Memorial nurse anesthetist who lived in Long Beach climbed into a pine tree, resulting in his whole body's being scraped and with a huge gash on his hand. Another employee spent six hours exposed to the storm's elements: relentless wind, water, tornadoes.

"People had the sense that if they showed up, we would take care of them," a hospital chaplain explained. "To many people in the community, this was their beacon. They had no other place to go."

They did not know where else to go. They lost their homes or where they were staying. They were looking for shelter, water, food, clothing, any kindness they might find.

Marchand realized Memorial could help. "Fortunately, we had an unfinished food court that we were able to dedicate to that shelter for them. It became known as The Village. As the aftermath continued, we were discharging patients that no longer had homes, no longer had caregivers, no longer had any form of transportation; so we discharged them to The Village and cared for them until we could get them to an appropriate shelter. Most of them were very shook up inside but seemed to come to grips. They needed comforting, emotional care and support, and the love that my nurses and other staff provided."

A hospital volunteer could offer water, juice, coffee, even sandwiches. "They just wanted somebody to talk to. They were scared. It was heart-wrenching to see people and their serious needs."

"Some were on oxygen at home and had run out. Their medicines were washed away; so we set up a makeshift pharmacy. Most of the pharmacies were destroyed or not open; so we provided—I don't know of any other hospital that did that," another nurse added.

While the hospital took in the wounded, the sick, the lost, Coroner Hargrove and Velasquez worked through the night in Biloxi to find and rescue survivors, returned to the EOC for about an hour, and went back to Biloxi. With still-working cell phone and radios, they also got calls from Long Beach and Pass Christian. "We would instruct them on how to make the recovery and how to handle the bodies," Hargrove recalled. "Once we were able to get that call through, I told MEMA we needed help from the medical examiner, and we needed the DMORT help from the north. I knew from the devastation we did not have the facilities or personnel we needed."

Disaster Mortuary Operational Response Teams—what Hargrove called DMORT—provide emergency help under the National Response Plan in recovery, identification, and services for individuals who die in a disaster. Ultimately, Hargrove and helpers in Hancock and Harrison Counties would handle 143 victims. But in those first days, in the first week after the hurricane, they simply responded, getting only one or two hours sleep here or there. Rescue first, recover later.

That second night, Hargrove had hardly settled when Velasquez ran from where she slept in the EOC to his office across the parking lot and street. "Gary! Gary—where are you?" she called. "We've got a major fire in a Pass Christian apartment complex that has not been searched."

Hargrove and Velasquez rushed to recruit the crew and fire truck—a pickup truck—that had come from Orange Beach, Alabama, with his friend and volunteer Sam Jackson.

"Then we got into my truck and followed it and the police escort all the way to Pass Christian to the fire," he said. "When we got there, we had the police chief and the fire chief—and the fire chief was overcome from having been exposed to the water during the surge and flooding, surviving that part of it. And he basically said, 'You're in charge.' 'Well you need to call them over here and tell them,' I told him. So I pretty much took charge of the fire scene at that point because of my fire background. I got things set up—got a front end loader and back hoe, basically commandeered that. The owner of it said, 'I'll operate it. Just tell me what you want.' I said, 'We need to cut a fire lane between us'—we were on Second Street—'between us and the beach because you have nothing but total debris on the coastline.'

"If that fire had escaped and gotten wind in it and picked up for any reason whatsoever," he explained, "we were going to lose a whole lot more than just that complex. So I got him doing that, and I got on the radio to the EOC and told them I needed bulldozers and air drops if they could get them. We did not have water; we were having to go pull water out of the ditches and streams with our fire truck and then transport it back in to use to put the fire out or hold it down."

Velasquez remembers the scene as "classic Gary Hargrove. We got there and saw two trucks, one Long Beach and one Pass Christian, and all these young firemen, just standing there because there's no water. We found a chief who was so frustrated, so beaten, so exhausted. And then you see Gary jump on top of this truck with his radio, and he's screaming, 'Send water! Send water!' And it's just a great image because he's got this tall, lanky body, holding his arm up with his fireman's jacket on and the radio, and he's screaming but can't get any feedback. And then he sees this guy with a bulldozer. . . And the firemen got some chainsaws and followed the bulldozer to cut that path. The bulldozer went around to make a trench so the fire wouldn't get to another building."

As others would during the disaster response, Velasquez looked to find humor in even the most dire situations. "When Gary and Sam Jackson and I got back together, I told them I want to know what about a natural disaster makes it okay for men to run around in their underwear! Not just the guy on the bulldozer, the sitting-down guy, but also this 108-year-old man from the neighborhood who came up with briefs blowing in the end where his butt used to be. And I'm sitting, and he's standing right in front of me, and he won't move. Then Gary comes up and is already laughing at me because I'm trying not to look!" Then her mood changed. "But could you imagine surviving Katrina and then a fire 40 hours later? Thankfully, we did not find any remain there; we want to think that nobody was left in that apartment complex, and we did not have any missing person from that address."

Chapter 16

People who worked with him would learn that Mike Beeman also looked for humor in the human condition, but when he introduced the Incident Command System to everybody in the Harrison County EOC conference room, he was as deadly serious as Katrina had been.

As federal liaison to the county and Spraggins' counterpart, he laid out the rules of the game: "Here's where the rubber hits the road. When you enter this room, there will be no griping, no complaining, no taking shots at each other. The only thing we expect to hear in here is what you accomplished in the past 12-hour period and what your challenges are for the next 12 hours. I need to hear this so we can figure out what we need to order, what we need to get from the federal government, from the state, from the EMAC (Emergency Management Assistance Compact). Everybody one-by-one will give a report from your role within the structure."

Beeman committed himself and the Harrison County responders to multiple meetings a day. On the hour from 7 am, specified individuals would hear situation reports, attend the planning meeting, take part in the command staff briefing, and/or go to a news media briefing. The command staff would re-group at 3:30, with the media briefing following at 4 and the executive staff meeting at 5. Situation report, planning, and command staff briefings would ensue at 7, 8, and 9 pm, respectively.

Commander Spraggins stepped in to introduce what he called the "hatchet theory," whose lesson they would also follow. "The hatchet theory is that when you go into a meeting, sitting there talking, and things aren't going your way, the first thing you do is to try to throw the blame. I told they could.

"Also, we had all kinds of jagged pieces of wood, metal, glass which was tremendously unsafe. We didn't need people getting cut and hurt. We had cut the gas and electricity; we had cut the grids so that wasn't a big deal. But the main thing I was concerned about was that we knew we had people missing. And some of those people were probably in that debris. And the worst thing in the world to me was for you to see your mother, your brother, your sister, or your child on CNN when they uncovered a piece of rubbish and there they lay. I didn't want that to happen. They needed to have the dignity of the family to be able to take care of their loved ones."

Mass media were also tightly controlled. "They could come in, but they could go only to certain points. They were very controlled. They couldn't run rampant up and down the Coast, nobody could run rampant—nobody. The only people who were allowed to go up and down along Highway 90 were the emergency personnel. Even homeowners who went in with identification had to go in and come out at a certain time.

"The whole Coast was under a curfew to start. We left it up to the mayors and the supervisors to tell us when they wanted to lift a curfew, and we took curfews and moved them when we needed. And it worked. We kept 24-hour curfew in Pass Christian and Long Beach for a long time, but Biloxi, Gulfport, and d'Iberville were ready to come off curfew sooner. We cleared and strategically lifted curfews according to what the mayors and supervisors wanted."

Response and recovery demanded workers, but curfews caused some distress for ESF-8 coordinators. Such tight traffic control prevented health care workers' movement. When Travnicek learned about this from his wife, Lora, in her nursing position at Garden Park Hospital, he turned to the Department of Public Safety and said, "You can save our lives. New Orleans is descending into chaos. We must prevent that here. Let the hospital personnel in—some of the people have been working 48 to 72 hours; they need a shift change." Within 20 minutes, anybody with a hospital identification badge could move into or from the security zone as needed.

"We changed according to what was open and operational because we wanted you to be able to get to work," Spraggins said. "Some people were able to work and the cities needed them there."

To facilitate their movement, the emergency management team developed a systematic use of identification badges. Each city as well as various companies, particularly utilities and other big employers, controlled its own, and all emergency team members had red badges—red to designate a badge of courage, the commander said.

The system originated and evolved for the Katrina response; today the system exists as a fundamental part of the county's emergency plan.

Chapter 17

As county health officer, with statutory authority and responsibility to protect the public health, Dr. Robert Travnicek took his seat in the Harrison County EOC before, during, and after Hurricane Katrina. In that job since 1990, he had talked at length about disaster preparedness and response with the county's former emergency services director, Wade Guice. Based on that tutelage and on-the-job experience, he was confident that he was where he was supposed to be. As the people's physician—surgeon general on the county level—he acted as family doctor to all citizens of the county.

Travnicek's seat adjoined that of AMR's Steve Delahousey or his representative. The two worked together as coordinators of Emergency Support Function-8 (ESF-8)—public health and medical services—under the Incident Command Structure.

"Our immediate problem the first five days after the storm was the lack of any form of basic sanitation and inability to meet basic human needs in the three counties along the Gulf Coast," Travnicek said. "Our sheltering system was destroyed, most only shells with roofs. No water, no sewer, no electricity. A couple of shelters that were suspect before the storm and opened against our advice now are only heaps of rubble. Most lost their windows."

Before they could attend to the thousands who had sheltered as their last resort, the two focused on individuals' health issues. The night of the storm, they asked public information officers to issue an important notice for dialysis patients. "All Tuesday dialysis patients from Harrison and Hancock

Counties should report to their respective units at 8 am tomorrow. Some of the units are damaged, but arrangements will be made for treatment. Any dialysis patient needing immediate medical attention should report to Memorial Hospital, where an acute dialysis team is present."

Another August 29 news release from the EOC advised that all patients and employees at Memorial Hospital were "unhurt and safe. Biloxi Regional and Memorial of Gulfport have the only two emergency departments that are functioning in Harrison County. The hospitals will not treat minor emergencies until the demand for their services declines. For minor emergencies, provide first aid at home."

After issuing the first water advisory that Monday—"use water sparingly. Be sure to boil or disinfect any water before drinking"—the public information staff soon switched to pleading with the mass media to "not report on ice and water distribution based on hearsay. Wait for official news release."

Talking points for morning and evening media briefings that Wednesday revealed Harrison County's new reality:

- Complete devastation for the entire county and the five municipalities
- Total loss of power and telephone communication
- No water
- Widespread gas leaks
- Not enough tetanus shots for patients
- Area hospitals are damaged and overcrowded
- Shelters are damaged and many are full to capacity
- A number of people are possibly still trapped in their homes
- There are fatalities

Twenty shelters remained open, but poor communications prevented Harrison County from providing "good figures on the number of occupants." The morning's hurriedly produced news document revealed important but disjointed facts. "Tetanus shot supply is low; Memorial in Gulfport is out of tetanus shots. Working hard to treat dialysis patients. Several successful rescues—some dramatic. Might still be people in need of rescue. Additional search and rescue teams from around the country are here to assist. As county coroner mentioned, multiple fatalities in Harrison County, but we are not prepared to release *any* number until all of the information is in, and we communicate with the Department of Health and the five municipalities. There have been some very inaccurate numbers given, and we will not speculate about possible fatalities. We are working with FEMA and MEMA to set up temporary housing for residents who lost their homes. Also discussing plans for debris removal and cleanup."

Compounding the problem was that nobody knew how many people had sought refuge in which shelters. Additionally, after the storm's landfall, shelter occupants reportedly moved among shelters in search of food, water, and better living conditions.

"It would be a week or so before Red Cross showed up and another week before they were organized enough to even give shelter counts," Travnicek said. "Fortunately, Harrison County had police and fire personnel with decades of experience in such disasters, and they went out in the middle of the storm to rescue occupants. Because of their flawless performance, we had no loss of life in a shelter as a result of the storm. But for the first four or five days it was total chaos."

Travnicek had been inside the EOC since Sunday. Only after utility companies "took the grid down" could anybody safely travel beyond the immediate vicinity of the courthouse. Managing the electrical and gas grids enabled safe travel through the devastated landscape. Shutting down the electrical plant also required hospitals, Travnicek's particular concern, to use emergency generators.

"Memorial Hospital of Gulfport burns 6,000 gallons of fuel a day, and they had around 24,000 gallons," he said. "Nobody ever anticipated anything like that. Trust me—we all got caught with our pants down to some extent. But I got to my car and drove up and down the thing, and I realize I can't actually drive down 90 because there's—not only are there lines every place, but portions of buildings. It's hard to know where the highway is; so I just went back in.

"And then the National Guard came out—and the Seabees. They were clearing stuff immediately, just clearing paths. We got a path to the Interstate—for about two weeks the Interstate itself was just a path."

Late Tuesday afternoon, he retrieved wife Lora from her job at Gulf Park Hospital, and they drove through a path along 28th Street. "You've got to remember the telephone poles, the debris—the Seabees went through there

with floodwaters and covered the environment. Nothing looked pretty, and the stench was "just unbearable," Senterfitt said. "Dr. Travnicek would tell us, 'Look. If we're not routinely washing ourselves with bleach solution, we're going to be in deep trouble here. If we don't take basic human cleanliness seriously, we're going to lose the whole thing, and that will be a big problem.' People were so overwhelmed by the scope of the disaster, at first they did not give him as much credit as he needed. When you're immersed in death and destruction, handwashing doesn't sound like a big deal. Even routine bowel movements don't sound like a big deal until about day three and day four, when we started losing people. Dr. Travnicek was making sure we had handwashing stations in front of the food line, that we had proper sanitation. That, beyond search and rescue, really became an issue as we got into the reality of this nasty natural disaster."

During those early news conferences, Travnicek appeared next to General Spraggins. "I usually spoke last. I knew immediately that he didn't know . . . he was careful with what he said. He was used to briefing Air Force squadrons; so he was comfortable in front of hundreds of people."

Travnicek himself also had been comfortable and confident, but stress from the storm and an ongoing conflict with what public health employees knew as "central office" unbalanced him slightly. An unsuccessful candidate for the state health officer position when the State Board of Health hired Brian Amy in 2002, Travnicek never established a collegial relationship with the Department of Health's new director. As medical doctors, both grudgingly gave the other professional recognition and common courtesy, but Travnicek gave little credence to the man he considered incompetent. Worried before Katrina about sufficient funding for public health— "Funding affects your ability to deliver services"—Travnicek after the storm would become much more concerned about whether the new director would allow him to continue to do his work or possibly fire him.

Sometime that morning—*Thursday?*—State Epidemiologist Dr. Mills McNeil appeared at Travnicek's desk in the EOC. "To those who knew him, or were around him even briefly, we knew he was prone to mood swings with bouts of shouting at people about things they could never have known or do anything about," Travnicek described. "That was in concert with his boss, State Health Officer Dr. Brian Amy. In short, Dr. McNeil was a self-serving, nasty little bastard. However, I knew he actually operated at a very high intellectual level and had a wealth of experience from around the world—I think due to a military background.

"He basically said in a brief private conversation, either at my desk in the EOC or in the morning briefing at the EOC conference room, that all of our shelters would soon suffer the same fate as New Orleans in terms of mass gastrointestinal illness, and I needed to do something about it. 'You've got 24 to 48 hours,' he said. I knew he was right. That was a not-so-veiled threat from Dr. McNeil as a message from Dr. Amy. Dr. Amy used this technique to threaten all his Central Office staff and the few health officers who had remained after Dr. Amy was hired and immediately announced he was closing one-half of the district offices.

"On a personal scale," Travnicek opined, "owing to the fact that I had been a candidate for his position, I seemed to be the most-often-attacked by Amy's minions like Dr. McNeil. No question, that was to please Dr. Amy, who became obsessed with my removal from my office and the department."

Travnicek described how he and many other public health proponents at the time evaluated the public health agency's state of affairs. "The department had been called the crown jewel of state government, but in 2003 when Dr. Amy finally got his hands on the department . . . within three months Dr. Amy destroyed the heart and soul of the Mississippi State Department of Health. He fired his best staff, all the while insulting and belittling lifelong dedicated employees into retiring or resigning. He kept a book on his desk in a prominent place: *Administration By Intimidation*. He made sure anyone who was unlucky or unwise enough to be in his office for any reason saw it. Dr. McNeil was his favorite.

"Dr. Amy would say to the recipients of this fate that he or she was disloyal and would relentlessly scream at them. Dr. McNeil was a carbon copy. To Dr. Amy I was the worst of the worst, the only person in the agency who had interviewed for the health officer position when he was hired and, therefore, a potential rival. He couldn't stand it. He wanted me out."

Even though Travnicek perceived that almost everybody else also considered McNeil "a raving lunatic," he kept quiet that day and determined to strategically use McNeil's information to protect Harrison County.

"I kept his dire prediction to myself, more or less, till the following morning briefing. Prior to the morning meeting, I privately informed Rupert Lacy, Bobby Weaver, and especially the two new but vital assignees, Marty Senterfitt and Chip Patterson from Jacksonville. Then at the briefing, I shocked the assembled, including General Spraggins, with the warning about sanitation and shelter conditions. I'd gotten word by courier—no telephone for a month—from Deborah Taylor, the infection prevention coordinator at Biloxi Regional Medical Center, that at least a few people from a particular shelter had presented themselves to the tents which acted as the emergency room and urgent care centers for the area served by Biloxi Regional Hospital. Their stools tested positive for norovirus.

"Deborah was a nurse I had trained in epidemiology during her nursing school experience," Travnicek said. "She had also been my epidemiology nurse at the health department some years earlier; so we had a long professional relationship, and I knew she was outstanding in her field. If she had told me water was flowing uphill, I would have believed her and acted on it."

Travnicek informed the command staff of the possible norovirus outbreak.

"He was The Great Predictor," Senterfitt said. "By day two or three he had told us, 'Guys, we have these shelters, and this many people in all these shelters with no sewage, no running water, and the toilets starting to overflow. And there's food being delivered to them, but they don't have any sanitation. Here's what's going to happen. People are getting raw sewage on their hands because they're out working—working during the day, trying to get the county back—but they're coming back and don't realize they've got raw sewage and other biologicals all over them and are eating and getting all that into their systems. In two to three days, we will get severe GI tract issues and severe intestinal issues, and we're going to have a crisis. It's going to build until we shut down.' Of course, everybody would listen, but there were so many more problems."

Travnicek said the only way to control the contagion was to close the shelter. "Evacuate everyone you can, immediately," he urged. "But they understood that was impossible. They were trapped. We were trapped. All of us understood that some of the horrible things happening in New Orleans were about to unfold here. We felt under immense pressure to avoid an epidemic, social unrest, riots such as had befallen New Orleans. I could see no way to avoid an epidemic. I think that was my lowest moment of the storm."

Chapter 18

Steve Delahousey took time to regroup. Uninterrupted adrenalin surges among a roomful of first responders for nearly a week began to wear on him and his colleagues.

Even though they could not stop, the natural numbers-cruncher sought and found comfort in survival from exploring the stats.

"Katrina was the most destructive natural disaster in US history," he reminded. "Katrina was not a powerful storm—it was big and slow-moving. It was a Category 3 when it made landfall in Mississippi, and that was the third landfall. It made landfall in Florida first, then Louisiana, and then moved across the Gulf to hit Mississippi. Twelve hours of sustained hurricane-force winds is what made it so bad. And that's what the National Weather Service told us: we've never had that before, that and a 28-feet-plus storm surge. We have that in the bulletin that National Weather Service issued at 10 pm on Friday before the storm. They predicted Biloxi would get a 28-feet-plus storm surge, and they called it correctly. And we were in the northeast quadrant, which is the most dangerous to be in, and the hurricane force winds extended 120 miles out from the center. Those five things really combined to make it the most destructive natural disaster in US history."

Responsible along with the county health officer for the public health and medical services in Harrison County, Delahousey continued to count.

"Harrison and Hancock County pre-storm had a population of about 250,000," he noted, "and 11 hospitals. Two evacuated pre-landfall, Gulfport VA Hospital and Keesler, which created a big problem; we usually depend on Keesler, the second largest military hospital in the world. For every previous hurricane I can recall, we depended on the military, and the military bugged out this time. As a matter of fact, we had to absorb their patients in the civilian hospitals."

Why Keesler closed did not surprise Delahousey. "They're right on the bay, at 10 feet, and a 28-foot surge would cover them. They knew they would get flooded."

Because power failed, authorities evacuated three more hospitals post-landfall, including Hancock Medical Center, "which kept operating through the storm even though the facility took on about six feet of water. Hancock Medical Center stayed open until the military shut them down. People were just walking up and wanting water, food, whatever we could provide. And then this North Carolina group came in and set up a Disaster Medical Assistance Team, DMAT, in the K-Mart parking lot, and they became known as K-Mart General. Hancock County had the best health care, post-Katrina, they've ever had; they had, essentially, a Level One Trauma Center. They had neurosurgeons and everything else Hancock County had never had. The services were in a tent, but the doctors were neurosurgeons, orthopedic surgeons, cardiothoracic surgeons—those guys were not *residents*. Those guys chartered their own jets to come down here. I took them around to the hospitals, and our guys said, 'Yes! We need some relief.' Those folks did a great job."

The two counties also had 15 skilled nursing facilities; four evacuated pre-landfall, a joint decision of the EOC and the homes' administrators. "And it was absolutely the right decision. All four were severely damaged or destroyed," Delahousey said.

He relayed more "interesting statistics—and nobody else has ever done this except me; no one's ever really been interested. Pre-storm, we had 1,700 hospital beds; post-storm, 935—we had a decrease of 45 percent. Also, 21 percent fewer skilled nursing facilities; home health patients, a five percent decrease; and special needs non-institutionalized patients—we don't know. And that's what bothers me. That continues to be my biggest fear nationwide: that this country is not taking care of our special needs population. They're not regulated, and we just don't know. There's no database. We don't know where they are, much less how many."

After Katrina's landfall, ambulances in the first two weeks handled a thousand inter-facility transports and another thousand pre-hospital transports. Despite media reports and public perception, the Gulf Coast experienced "not a lot of trauma," the veteran EMT-Paramedic said. "Number one, a lot of people left. And even though you hear about people coming down here and saving all these lives, the call volume was comparatively slow. Moving people, getting them out, and taking them to other hospitals—that was the issue. It irked me that a couple days after the storm you see on TV that the Red Cross was begging people to donate blood. We didn't have any *trauma* down here. The Red Cross always needs blood, but we did not have trauma victims because of the storm."

He counted 65 supplemental mutual aid ambulances delivered to Harrison County pre-storm and another 60 after the storm. "I had 13 ambulances damaged or destroyed," he said. "That was one of the most frustrating things when we started restoring communications. My bosses wanted to know, 'Is everybody OK?' and I didn't know. I had 13 ambulances, 10 crews were unaccounted for—what a helpless feeling when you don't know if they're dead or alive. We ended up losing no one, but we didn't know for days. We had put them in safe places by Camille standards. We'd put them in fire houses, hospitals, near police stations, schools, churches—safe by Camille standards. Camille was the benchmark. Who'd ever have thought we'd have anything worse than Camille, a Category 5 storm? Now we'll use Katrina."

Peak ambulance deployment increased from 21 before the storm to 33 just four months later, 57 percent more. "Thirty percent of our population was gone, but we had to have more ambulances because the bridges and roads were out," he explained. "We have strict response time standards but under a state of emergency, we are not obligated to meet those. But I said that we were going to continue to make our response times; so we kept enough ambulances on the road post-storm to make sure that we did. We had 57 percent more ambulances to transport nine percent fewer patients. That underscores the importance of infrastructure—it all ties together.

"Probably the most important statistic in my book is that we had no hurricane-related death in Mississippi health care facilities," he affirmed. "None. We had no death resulting from patient evacuation."

Before the Gulf Coast of Mississippi had any inkling of Katrina's

- Safety and security of human drugs, biologics, medical devices, veterinary drugs, etc.
- Blood products and services
- Food safety and security
- Agriculture feed safety and security
- Worker health and safety
- All hazard consultation and technical assistance and support
- Mental health and substance abuse care
- Public health and medical information
- Vector control
- Potable water/wastewater and solid waste disposal, and other environmental health issues
- Victim identification/mortuary services
- Veterinary services
- Federal public health and medical assistance consists of medical materiel, personnel, and technical assistance.

Experience as emergency planners and responders to previous hurricanes had prepared Travnicek and Delahousey to think about and/or respond to most of those functional areas—certainly those they and others in Harrison County EOC recognized as particularly important to Harrison County. As the Katrina story evolved—and Delahousey reminds, "We were actually under a State of Emergency from August 2005 until January 2008"—they personally dealt with sewage and wastewater disposal, disease surveillance, food and potable water, disease control, and restoration of the health and medical infrastructure.

The ESF-8 twosome in Harrison County worked with the local health and medical infrastructure before, during, and after the storm. They directed emergency response action for such health facilities as hospitals and nursing homes. They monitored patient movement and tried to assure that all citizens, especially those with special medical needs, were evacuated or as safely housed as possible. They coordinated with all other emergency support functions represented in the EOC.

"Steve and I agreed that as county health officer I could claim authority, even though I might not have any specific," Travnicek said. "Dr. Amy and Jim Craig and Art Sharpe told me that public health had nothing to do with the emergency medical care and private medicine, but I've done everything a health department does except deliver direct medical services to people who are ill."

Travnicek knew his job covered, at the most basic scope, emergency medical care, preventive medicine, patient movement, and a whole lot of environmental health science. His range of responsibility would cover food, sanitation, water, sewage disposal, and condemnation of tons of hazardous materials, largely spillage from the Port of Gulfport, where hundreds of shipping containers (the kind that fit on railway cars) sloshed ashore, their contents scattered throughout the landscape.

Accordingly, early into the aftermath, he asked EOC Commander Spraggins to call together the county's attorneys, judges, and local court officials. "They weren't familiar with public health laws, and I needed them to familiarize themselves with public health laws because I was doing some major-league shit, having whole buildings bulldozed down without ever seeing them because Amy wouldn't give me anybody to go out and see it—I hadn't seen it. And he wanted to send the stuff up there to sign, but he didn't have any authorization. If we would have done what Amy and those morons wanted to do, the shrimp would still be there. I'm the county health officer, and I knew I would need legal backup and support."

Chapter 19

Sanitation, food and water, and fuel became the most highly desired yet least available commodities in Harrison County for up to two weeks after Katrina's landfall.

The first formal situation report, or sit rep, recorded for the EOC on September 5, a week after the storm landed, named those as top priorities for disaster responders. They would remain at the top of the list well beyond the disaster's second week. Public health and safety experts assumed everything the surge had touched to be dirty, unclean, insanitary. Water became a toxic brew as it mingled with raw sewage, dead bodies, leaks from oil and gas containers and lines, and such environmental hazards as asbestos and other dangerous minerals. Undoubtedly, the water contained household and industrial chemicals, contamination from sloshed containers of everything from lawnmower gas to medical wastes, and bacteriological entities.

Professional Engineer Kamran Pahlavan headed Harrison County Utility Authority. After Katrina landed, he could count only the losses. His county had lost communication, utility poles and wires, trees, the contents of his flooded office and its vault, and many of the 34 pump stations used to move sewage to treatment plants. Because of flooding and, in one instance, a house's being deposited into a plant, even the wastewater treatment facilities were considered lost.

"Obviously, we had to do something," Pahlavan said. "For the first time, the sewer guys had become important! Nobody could do anything until we were able to remove the sewage, to restore the water, to get power."

Sewage in the streets, salt water in the generators, pump stations under water, hot August days and nights, no electricity, high humidity—and initially Pahlavan could not locate most of his workers. As they did return to work, he negotiated extra help from his contractor and gratefully accepted disaster assistance from EMAC crews.

"We went back at least 200 years right after the storm," he said. "We had to use the old technology. Our system actually went under water so we could not chlorinate; what we had to do was the minimum, the old-fashioned way. We actually had these chlorine tablets we just floated in; it was not a perfect job, but we killed a lot of bacteria before it got back into the water, back into the bay.

"Everybody thinks everything happened to New Orleans. I feel bad for those folks—I really do. But when you look at the magnitude and compare, it's actually day and night. *Here* it was night. The weight of the water that came in—particularly in Pass Christian—was so massive that it destroyed part of the underground. It destroyed infrastructure, the sewer lines. Katrina came in and completely took away. Mansions are gone. Nice buildings—nothing left. Where there was steel, nothing is left."

§

District Health Officer Dr. Travnicek also saw the gravity of the sanitation issue. "There were no sanitary facilities, no sewers, no electricity except for hospitals, correctional facilities, and emergency departments, and they were running on generators. People were relieving themselves on the streets and behind bushes, and a few in the shelter halls, too, according to police reports. There was no water; there was no sewer for 10 to 14 days. Downtown Gulfport had been under water, flooded, and all the sewer system's lift stations failed. People were shitting in the streets. I needed three things: I needed potable water, I needed an unlimited supply of safe water, and I needed the sewers to run.

"There was no water. It was all bottled water. And then we had a breakdown about day five—we had norovirus in one of our shelters, and I closed the shelter and ordered everyone evacuated."

While he had to rely on outside help to deal with contagion in the shelters, Travnicek single-handedly managed at least one policy issue.

"I was able to convince them sewers before water, sewers before water, sewers before water—*gosh*. I had to fight that for two weeks," he nodded affirmatively. He remembered one of the classic stories—Orlando suffered a minor hurricane and lost all electricity but not water; people kept flushing and put downtown under two feet of sewage. "I must have gone to a hundred meetings over that one issue—it's minute by minute. The mayor of Gulfport was going to turn on the water, but we were able to get to him . . . It's an emergency when you're in the operating room. I practiced medicine for 22 years before I ever did public health—I probably did 200 procedures. When you're in the operating room, you don't give a damn whether or not your office is on fire. You just don't care because it's desperate. This was desperate. We absolutely had to get the sewer system operational before we turned the water on; otherwise we would have flooded everything a second time with untreated sewage.

"We were at the point where it's survival of a culture here. It became clear about the second day that this Gulf Coast might not survive. We had 400,000 dead chickens everywhere. We had two million or four million pounds of pork bellies strewn all the way from Gulfport to damn near Pass Christian. I mean, this was desperate."

Getting the sewer system back took top rank among immediate priorities: "The first time I saw the mayor of Long Beach, he told me the city south of the tracks had ceased to exist. I didn't believe him, but I factored it in and asked him about the water pressure. He said they'd maintained 20 pounds—'Keep it up,' I told him. 'Dump the water, but keep the pressure up.' We had to get power to the lift stations. We had to get generators to run them, hauling generators from lift station to lift station, effecting a series of dumps, moving sewage one lift station at a time. And then we finally were able to restore electricity. Even then, for months we had no choice but to run it out to the ocean."

As ESF-8 coordinator, Travnicek was confident that hospitals and even shelters knew how to meet basic, personal sanitation standards, but he also knew that nobody enjoyed using portable toilets, much less red plastic bags.

Biloxi Regional Medical Center's Deborah Taylor, infection control nurse, explained: "We had no water. No toilet. But I do have a policy on how to use a red bag! People need to understand: people still have to go to the bathroom. We actually ended up having to lock the bathrooms—we

Long Beach and that smell came back. It's still in those rotten buildings."

Atterbury described "thousands and thousands of tons of pork bellies and chicken and seafood" destroyed in shipping containers at the 200-acre Port of Gulport. "We had large shrimp boats that needed to be moved. We had thousands of vehicles. Not only FEMA but also EPA, MDEQ, MSDH—so many federal and state agencies shared responsibility for cleanup. It was just more than any one entity could handle.

"I was working the EOC one night, about 1:30 in the morning, and this man comes in and says, 'M'am, I've got a little problem.' And I said, 'What problem?' and he said, 'Spoiled seafood.' This was two or three months into it. He said, 'I'm moving stuff down in the Copa (barge casino). We finally got to where the freezers are.' I said, 'Oh. How much are we talking about?' And he said, 'About 10 tons.'

"We had shrimp and seafood factories here, and those are things you've got to deal with. . . You can't just pick stuff up, throw it into a pile, and hope somebody else will take care of it."

§

At MEMA headquarters in Jackson for Katrina's landfall but on the ground in Harrison and Hancock Counties soon after, Federal Coordinating Officer Bill Carwile with FEMA and MEMA Executive Director Robert Latham struggled with "managing shortages. If you need 200 truckloads of water today, and you get 10, the Unified Command Group basically had to decide who's going to get it today. Rationing—that's what we were doing! We were managing shortages."

Hospitals, the shelters, and the homeless felt those shortages, too. Even though the EOC had told evacuees to the shelters to bring food and other essentials for three days, most did not. Nobody had planned to become homeless. Hospitals had emergency plans in place, but none had foreseen the influx of people who came simply because they had nowhere else to go. Food and water supplies disappeared fast.

FEMA Mike's first question up the chain of command dealt with when commodities would become available. He had assured Harrison County commanders that he could effectively battle for their common requirements, a shared need which surely included food and water.

About the third day, an unfortunate set of circumstances dropped into Bill Carwile's lap. Beeman explained: "All the commodities we were expecting to come to Mississippi—all the water and ice and other commodities—the decision was made at the national level to send it all to Louisiana. So nothing was going to come in to Mississippi. Nobody ever knew that happened because we turned around—and Bill told me he had already been on top of it. I do know that when I talked to my IMT team from Jacksonville the morning of the third day, they had told me the State of Florida had been approved for Category A and B reimbursement (emergency protective measures) from the federal government. Food and water come out of those two categories. So I asked them to contact the State of Florida and ask them how quickly we can get food and water and other commodities headed in our direction. They'll be reimbursed for them. So the guys went off for a couple hours and came back and said, 'They can have a hundred truckloads headed in our direction within 24 hours.' And I said, 'Head it on!' Nobody ever knew that happened because the commodity flow continued by calling on the State of Florida; they just kept it coming. And that's the way for it to happen: transparent to anyone else when you make the system work within the structure. It seems simple, but until you're in the throes of finding food and water for 184,000 people. Now, granted, a lot of them had left, but there was still a large amount of the population there."

In addition to those truckloads from Florida, individual charitable organizations also began to deliver food. Harrison County, whose disaster preparedness plan called for establishing five points of distribution (PODs), opened 23 PODs that first week—and would have opened more if sufficient resources for staffing had been available, Logistics Officer Lacy said. Outback Steakhouse set up just outside the courthouse to feed emergency responders. The Church of Jesus Christ of Latter-day Saints sponsored a hurricane relief center in Biloxi to provide victims supplies of water, hygiene items, canned food, and baby supplies. By Thursday, the Salvation Army, Red Cross, First Baptist Church, Cornerstone United Methodist Church from Naples, Florida, and other charitable organizations had opened feeding and food delivery sites. Tens of thousands of people whose homes were destroyed or gone desperately stood in lines in the heat to get food and water. Public information officers issued news releases that identified locations for the mass feeding sites, canteens, and mobile food delivery vehicles; they promised 2,500 port-a-potties by Saturday. A release on Saturday announced that PODs would be "lean" on water and ice—some might even deplete their supply—but that 75 tractor-trailer loads of water and ice per day from the State of Florida would increase to 100 trucks per day.

Working to assure that everybody in the county could get at least one meal a day, the command staff thought they were doing OK. "Two meals, we were ecstatic," Senterfitt said, "and I'm talking just basic hand food. But on day four or five, we found out that about half of our food trucks aren't rolling out to deliver meals. 'What's the problem? Do you not have enough food?' 'No—we have enough food, but we can't go because health code says the trucks with greywater tanks—dirty dish water—cannot deliver food if greywater tanks are full.' No company was available to service the tanks. So they were sitting in the truck and waiting for somebody to come empty the tanks appropriately.

"We brought the problem to Dr. Travnicek. 'Wait a minute. This isn't sewage; it's grey water—dirty dish-soap water versus sewage and dead bodies. We've got sewage and dead bodies, why are we worried about greywater?' Well, we've got a regulation, an ordinance, that requires proper disposal. We're not meeting the common sense threshold here. I went over to Dr. Travnicek and said, 'Hey, look. Here's the problem,' and I laid it out. He said, 'That's ridiculous; food is more important. The dishwater's better than raw sewage on the ground.'"

In short order, a laptop computer and printer produced a one-page letter that cited the crisis and allowed response trucks to dump grey water into storm sewers. Both the county health officer and incident commander signed, allowing the trucks to pull their levers, dump dishwater into the drains, and start feeding again. Senterfitt remembered, "Dr. Travnicek laughed about it. 'Number one, we're cleaning the storm drains, and number two, we're getting food out to the people.' Those decisions that were having to be made on the spur of the moment made all the difference."

§

Mississippi news teams quoted Governor Haley Barbour's saying that

Katrina "was more than a calamity for the Coast. A lot of Mississippi was clobbered." Most of the state remained without power and under boil water notices. Schools and businesses remained closed. Lack of power fueled a demand for batteries and generators. Scarcity of gasoline resulted in long lines of would-be buyers whose tensions tightened, and at least one person died by gunshot after quarreling over a bag of ice. Katrina caused a whole lot more than inconvenience.

Katrina also "crippled substantial portions of the country's energy infrastructure," *The New York Times* reported on September 1, 2005. "In Louisiana, Mississippi and Alabama, electrical power was out, refineries were drowned, and most of the offshore production of oil and gas had not resumed . . . Throughout the South and Midwest, service stations were beginning to experience some shortages in areas where gasoline is usually transported by pipeline from the gulf. Nearly two million barrels a day of refining capacity has been knocked out by the storm and could take weeks to return. . . More than 90 percent of the gulf's daily oil output - or 1.37 million barrels - remained closed yesterday, while natural gas production was down by 83 percent, or 8.3 billion cubic feet, according to the Minerals Management Service, a unit of the Department of Interior. Coast Guard crews reported that up to 20 rigs and platforms had either sunk or were adrift, Larry Chambers, a public information officer, said. At least one gas rig has caught fire. The Mars platform of Royal Dutch Shell - which alone accounts for 15 percent of the gulf's oil production - is 'severely damaged,' the Coast Guard said in a release."

At Mississippi State Department of Health, Jim Craig worried about managing fuel shortages. "I still remember when a reporter from CNN stuck a microphone right in my face and asked, 'What's the largest problem you have for health response for Katrina?' And I said, 'Fuel. I've got people in the hospitals on generators, and any minute those generators are going to run out of fuel; so they're asking us to send resources down to evacuate them, and I've got to put fuel in the ambulances to go down and get them. . . so if you ask my leading problem, it's fuel.'"

In Gulfport, Harrison County Operations Chief Bobby Weaver learned just days after Katrina's landing that fuel—or the lack of fuel—had become a life-or-death issue.

Weaver recalls the "harrowing circumstance" of getting a telephone call from Memorial Hospital. "They informed me they were down to two hours of fuel for the emergency generator, and if they couldn't get fuel for the generator power, people could die."

Memorial already had lost power for a short time about 2 am August 30. "And I promise you 20 minutes can become a long time," a Memorial executive remembered. Said Jennifer Dumal, vice president for patient care and also chief nurse, "In an internal building, when it is dark, it is really dark. And when you know you've got a surgical case going and you're trying to get to your patients on ventilators—we had NICU babies, we had an ICU, and we had a surgery underway—when we knew we had to get to those places, whoever you are, it seems like it takes you an eternity to get there!

"They did that surgery by flashlight," she added. "A person 'doing what was she was supposed to do' appeared with flashlights. . ." and in the next revision, the hospital's emergency preparedness plan called for strategic use of both flashlights and out-of-area cell phones.

That temporary blackout occurred, VP for Administrative Services Larry Henderson said, because a fuel relay switch failed in a standby generator. Workers could repair that, but where could they get more fuel for the generators? The individual who called the EOC for help had no idea who usually supplied their fuel, and by then no communication system remained functional. ESF-8 coordinators in the EOC tried to help, but even a preventive medicine physician with special training in public health could not help.

"At this time," said Dr. Travnicek, "all of us at the EOC had only emergency power—no air conditioning, no computer, only emergency lighting and fans. Things were really bad in our bunker on the first floor of the courthouse. Each hospital is a different story. In that big room, the hundred people there were thinking about and trying to deal with roads, sewers, water, hospitals. I had told all the hospitals for 10 years to have fail-proof, independent generators and as much diesel fuel as they could get, free-standing commercial water wells, and bullet-proof windows. All hospitals lost part of their roofs. Water, walls, floors, emergency departments—problems everywhere. But I told them they could not leave; they could not abandon their patients, staff, and missions. After the storm, none of them could get additional water for three days. FEMA had only 56 gallons of diesel fuel; so we spent hours trying to get fuel to keep Memorial open. And Memorial burned six thousand gallons a day. I'd told all of them they had to be self-sustaining; so it was real emotional for me. I would not allow them to close. And that's when Gary Marchand became a mixed hero. He got fuel. Faced that close with the risk of going down, Marchand got fuel. And Memorial took care of its people."

IMT Deputy Commander Senterfitt from Florida saw the incident with a different set of eyes. Describing the situation as "heartbreaking," he heard the health officer reveal in the sit reps meeting that Memorial needed help: "People are on life support."

That day, he said, "We sat down at the computer and ordered the fuel, said we needed it for the hospital. . . Later that evening we got the message back from Florida that the fuel truck was on the way and would be there in the middle of the night. When you're ordering 1,200 to 1,400 orders a day—we didn't have time to physically watch each order. So we ordered the fuel truck. It was on the way. Then we took our eyes off that and watched another problem. The next morning during the operational briefing—I was actually standing in the front of the room talking—somebody from the medical side came in upset, very irate, and he yelled across the room, 'You lied to me. You told me we were going to have a fuel truck here this morning—well, there's no fuel truck, and I hope you can live with yourself because we've got people dying at the hospital now because we have no fuel.' Of course, I'm knocked backwards. Chip Patterson, who was on the other end of the room, looked at me and gave me a hand-motion to indicate 'keep going with the meeting,' and Chip ran out the door with this guy. What we found out later was that, unfortunately, the truck did not arrive . . . We found out later that, kind of a consequence of the overall disaster, the fuel truck that had been ordered and was en route to the hospital actually got commandeered by law enforcement when it came into the area; they said, 'We've got more of a need,' and they didn't check but took it off to do another mission. Because there was not good coordination among all the entities, that fuel truck for the hospital never arrived. A little bit later that day, we were able to divert fuel and get it to the hospital.

"In times of a disaster, you cannot have a situation where everybody starts stealing from somebody else," Senterfitt verbalized another lesson learned. "Somebody thought, 'Hey, we need fuel for...' whatever our mission is, but they don't understand that the fuel truck they're commandeering already had an important mission assigned that was just as important, or maybe more important than their own mission. So a big part of what we were trying to do these first nine days was just bring order to chaos. Get everybody organized. Make sure nobody is stealing resources from each other, make sure that the resources are going to the right place and in the right time with the right priority."

The immediate problem resolved, the command team took action to deal with the fuel issue in a way that would meet the county's comprehensive needs.

FEMA Mike knew fuel supplies were "critically low for emergency response vehicles as well as for generators used to power a wide range of required facilities and functions, including hospitals, countywide shelters and feeding sites, and even refrigerated trucks holding human remains." He and Operations Chief Bobby Weaver tackled the problem.

"We had incoming fuel that was being confiscated," Weaver said. "We couldn't prove who it was by, but when you've got every agency in this state needing fuel to provide service, and fuel trucks are coming into Harrison County, but gas was not coming to Gulfport. . . We said, 'We've got to come up with a fuel plan. We've got to get some fuel for ourselves.' So we sent one of my Sand Beach Authority employees and a representative from the IMT-Florida to Pascagoula, to the Chevron Refinery. We said, 'Go over there; knock on the gate; do what you have to do. See somebody. Get some fuel over here.' So, arrangements were made. They had fuel, but they had no way of trucking it."

Commander Spraggins filled in the story. "We had one gentleman that walked in—he's a Mississippi hero, too. Mike Matthews. He owned a little oil distribution company just north of the EOC. Mike and I were friends, and he walked into the EOC the day after the storm and said, 'Joe, I don't have a whole lot, but is there something I can do to help you?' And I said, 'Well, I don't know—what have you got?' And he said he had some trucks but no fuel, that if we could find some fuel, he could go get it and bring it in."

With agreement and hand shakes, they sealed the deal. "No contract," Spraggins said. "Just word of mouth and maybe a sheet of paper that he and Pam [Ulrich, finance chief], sat down and worked out together. I don't know how many millions of dollars worth of fuel that man hauled—no telling how many gallons of fuel he hauled, but a lot. Without a guarantee of payment—he just said, 'We'll do it, and you can pay me when you can.'"

Adding Sand Beach Authority's fuel truck and another out-of-state volunteer responder to the deal assured full resolution of the fuel problem. "It just so happened that a gentleman came up to me—a law enforcement officer, I think, maybe from Indiana—and I don't know if he was from a city or a state trooper," Weaver said. "He reported in and said he was sent out here for Bobby Weaver to tell him what we needed him to do. I had two choices: one, I could send him out to the sheriff, or, two, I could put him in charge of running escorts on our fuel supply. I said, 'Well, we're having fuel being confiscated. If anybody confiscates any more, they're going to have to confiscate a patrol car as well.' So we assigned that unit for 10 days to run fuel escorts for our trucks between here and Pascagoula. I ran my truck 24/7; I assigned two people to it for a 12-hour shift and another two people for the next 12 hours. And we ran fuel."

Chapter 20

Tina Stewart with her mom and the rest of her family got back to Gulfport Wednesday afternoon; her then-husband and older son had returned Tuesday. Because their house occupied what they learned was the "highest dirt in Bayou View," their home did not take on water. From there, they took everybody to Picayune, to her parents' home. A nurse, she went back the next day to the courthouse in Gulfport to learn where she might volunteer to help with the disaster response.

"The girl at the table said, 'You need to follow him' and pointed to Travnicek. I had never met him, the mad scientist, but I followed him all the way back into the EOC operations room. He sat down in his chair then looked up at me, and I said, 'I'm here to volunteer.' 'Great,' he said, 'what can you do?' And I asked, 'What do you need done?'"

Amidst the chaos and with that exchange, Dr. Travnicek told his new helper about information from the shelters that some people who were diabetic did not have their insulin. "Basically, he needed an assessment of the shelters, to know what their needs were. So they lined me up with a county car, told me where the shelters were, and off I went—not knowing what the hell I was getting into."

Five or six shelters required attention, she learned. That afternoon and the next day, she visited Biloxi High School, Orange Grove Elementary, and a site in Long Beach. The second day almost provided an opportunity for her to deliver a baby, but then she "heard sirens, and a fire truck drove up, and I'd never been so glad to see them in all my life!"

The census and conditions varied among shelters she saw. Staff seemed to have tried to separate evacuees into groups by their needs, and one site had even set up a small clinic. Both staff and residents were "generally cooperative, certainly not hostile, and eager to tell what they had been through," Stewart said.

Schools, primarily, served as shelter sites. Since his term in office started in 1980, Harrison County Superintendent of Education Henry Arledge had made sure that all principals and administrators throughout the five districts were trained in Red Cross and FEMA shelter management.

"The school houses are about the only places that most communities have available as shelters," he said, "and the schools are not very well built for shelters. But it's the best we have. We've got millions of dollars invested in facilities, equipment, and supplies. We opened the shelters, starting Sunday. Our employees have to stay at the shelter; we have schedules, and we have somebody there representing the school district. We turn it over to the Red Cross, if possible, but they generally don't have volunteers. When people come in from Houston, Dallas, wherever, it's hard for them to manage. Our principals and administrators in this district do an outstanding job."

Arledge recalled opening 15 shelters and managing them for three to five days before consolidating to just six shelters. "We kept those open for about 21 to 23 days under horrible, deplorable conditions: no restrooms, no water, no electricity . . . and then, as time went on, some things started coming back.

"We had to referee a few little fights," he admitted. "My staff—with, after four or five days, with no food, sleep, rest. People get tired and irritable. They thought another school district was sending people out to us; so I had to referee. You have all kinds of problems and issues. We had a man who died in one shelter, in the middle of the storm. You just have to do the best you can. The principal put him into a body bag and shut his eyelids. At another principal's school—a deputy got a call that a police substation in d'Iberville had five deputy sheriffs stranded—and a couple dogs. He took a school bus and drove it to the water, and they swam out to the bus, and he backed out to save lives.

"You have two different types of shelters," Arledge explained. "You have to have a shelter for people coming in from the hospital, 'special needs' people, and you have to have shelters for just regular people. Generally speaking, people are nervous but very cordial and work together in the shelters. We had no Superdome event, but we had police officers there, and they knew if something was fixing to happen. We had a few problems with some kids getting into and tearing up a teacher's stuff. They had a camera in there and took pictures of themselves and left the evidence . . . Then we started figuring out how to get school started."

Schools in Harrison County would not start classes that fall until October. For those final hot days of August and well into September, many of those settings normally devoted to children and learning served as a refuge of last resort for now homeless citizens.

There largely because they had no other place to go and no way to get there, the individuals and families in the shelters had no way to know what constituted conditions beyond their own. Since arriving on Sunday and Katrina's landfall on Monday, they had been cut off from everything else in Harrison County and the rest of the world. They knew only that they had survived a surreal event, that they were hungry and thirsty and miserable, and that they needed help simply to survive. Those who saw Tina Stewart arrive wondered if she had brought that help.

"I pulled up at one place, and I had my supplies and started getting out, and they started crowding around me," Stewart described, "and I wondered, what am I getting myself into? People wanted to know why I was there, what was I bringing, and could I help them. It was amazing, but all of them were very calm, nothing out of the ordinary. They were not rowdy. There was one guy who was barefoot; he had on a pair of hospital scrubs and looked like he had been beaten with a baseball bat. He'd been in a tree for however-many hours and floating in water and beaten up by debris; he was still very much in a state of disbelief.

"The storm had left no social barrier, no financial discrepancies, and everywhere you went, people who before the storm had everything all of a sudden had nothing. It didn't matter who you were before, on Sunday. You were homeless on Monday.

"I think everybody was in the mood to help," she hesitated, "but out there it was very different. A lot of people who were at the shelter, running the shelter, were fearful—not debilitating fear, but what are we going to do? How are we going to feed these people? What are we going to do next?"

Unable to provide answers, she returned to the EOC and reported what she saw; Stewart realized that she had observed and brought back to the ESF-8 leader important information. She knew that her training as a nurse and supplies that enabled her giving some individuals breathing treatments were valuable, but "that wasn't *the* issue. The places needed port-a-potties. There were bathrooms that had stools that they had bagged and reports of people cleaning them out with their hands—you have a pregnant wife who has to go to the bathroom, you know? It was horrible! Deplorable."

Back at the EOC, answering telephone calls, responding to requests for help within the ICS, she witnessed disparities in distribution of portable toilets and medical supplies. She talked to people who wanted to donate supplies and their services—doctors and nurses who wanted to volunteer. Clinics were going up; organizations were coming into the area and setting up clinics. One pharmacist reported his store too damaged to open and that he wanted to donate all the non-narcotic drugs, antibiotics, first aid and other over-the-counter medicines.

For 15 days, Travnicek considered the volunteer nurse his one hope. Especially through those first two weeks, more and more calls to the EOC for help involved environmental or medical issues. AMR representatives easily handled the medical transport and hospital issues, but Travnicek had no public health help. Watching the AMR chair shift among three or four occupants during the 24-hour days, he juggled more requests for sanitation and shelters help. "Normal medical training does not prepare the average physician for any kind of disaster response," he said. "My ability to cope with mass trauma in a long-term event came largely from my decades in private practice, including 22 years working in a rural hospital and managing my own emergency room. I was able to work for extended periods of time with brief snooze breaks and go for days to weeks."

§

From the time she entered the EOC each morning until she left at the end of the day, Tina Stewart fought to handle all the phone calls. Travnicek had told her that the line at their desk was one of five secure, buried lines into the EOC. On that one telephone, she could receive or make calls from or to anyone, anywhere.

"I couldn't answer a number of questions," she acknowledged, "and some of the time those calls came from Jackson. They would ask, 'Now, who are you?'" Those calls came from the State Department of Health in Jackson.

"Yes, I picked up on friction. It became clear—I was only there for 15 days, but in that 15 days, it was very evident that Jackson was trying to manage the disaster on the Coast from Jackson, and, yes, a few people were here but they weren't at the EOC—they were up on Dedeaux Road. It was very disheartening to know that they had a district director in place, here, and it was those guys working against this one. Travnicek had identified potential problems—like the chickens, and the shrimp, and things that were going to be huge health issues; nobody really seemed to care a lot about that. Basically, I think he felt like it didn't really matter what he did; he was going to piss them off. And finally, it was like, 'Screw it. I'll just do what I know needs to be done and do what I know how to do.'"

§

Travnicek knew his county had been "bombed back to the Stone Age," that too many too sick, too hungry, too thirsty, and too frightened people were crammed together in inadequate shelters, and he knew how to accept help. Since 7 am that Wednesday, he had attended meetings—situation reports, planning, command staff, and even the executive committee meeting.

"By Wednesday, Joe Spraggins had been on the job for about 96 hours," the ESF-8 coordinator reckoned. "He found himself in the exact chair occupied by the great and legendary Wade Guice. Thank God, General Spraggins was as incident commander outstanding. Fortunately, if it weren't for a strange set of events, and if the following sequence of events did not play out exactly as they did, we would have looked like New Orleans. We would have been exactly in the same place they then occupied in the national press, with the need to answer a lot of hard questions, including our judgment and competence.

"The protocol demanded that command was to hand off every 12 hours to another chief of operations. This is something General Spraggins never did, and after a very discouraging report from the 7 pm briefing Wednesday, he went to bed and told his senior staff not to wake him. Linda Atterbury was actually designated the deputy chief on the organization chart; Linda had forehead, bowed her head, and prayed, "Oh, my God." On her third or fourth trip—*who could count?*—to rescue more homeless Mississippians from what she called "hellholes," she wondered how and why so many could have stayed to face such horror.

Before Katrina slammed Mississippi and left incredible destruction all along the coastline from the Florida Panhandle halfway to the Louisiana-Texas border, Bloodworth feared the monster storm with her family, sheltered safely at St. Lawrence Catholic Church in Fairhope, Alabama. After the storm moved northward, she realized that her parents and two children would be okay. She also knew her career of working for non-profit organizations had perfectly prepared her to help other families.

"When we realized how bad the storm was, a few of us loaded up on the church bus and started driving west until we began seeing people on the roads looking dazed and traumatized. We found a building with people huddled, injured, wet, and scared. We took as many back to Alabama as we could fit."

Wind gusts of up to 80 miles an hour near Mobile and heavy rain of three to six inches along the 95-mile stretch between Fairhope and Harrison County left even Interstate 10 littered with debris. The normally hour-and-a-half trip seemed much longer, and they got only so far as Moss Point.

Bloodworth and her fellow parishioners first saw people along I-10.

"We got to Moss Point, and we saw people injured and in shock, including police officers. They didn't have control of any situation. So we drove around and found our first shelter. People were trapped inside, and some of them were injured, and a lot of them had lost their homes, had swum, had gotten to the shelter somehow and been there overnight, with no water, no electricity."

The Catholics persuaded two busloads to leave and promised to come back for more. Bloodworth reached out to a couple she knew who owned a bus company in Daphne, Alabama. "They knew if they donated the buses to us, they would forego the FEMA contracts that would inevitably come," she said, "but they donated three buses anyway. They went bankrupt later, and that still makes me very sad. I put out calls on day one to everyone I knew that had medical expertise of any kind and loaded them up, too."

On the third of four trips, the Alabamians got to the high school in Biloxi.

"It had been a full 48 hours. They had been stuck in that school with no food, no water," she remembered. "People who were injured had wandered to this place and already had open wounds and infection setting in. And there was definitely a terrible bacteria going on in that school. Nobody at the EOC wanted me to say it. I brought a doctor in with me, and they yelled at her and told her not to say it. They were intimidating—not Dr. Travnicek but other EOC officials. I do remember Steve. He fought us and told us we could not take the victims out. He told me that it was against the law. As I recall, it was Dr. Travnicek who helped."

Dr. Darren Scroggie, a rheumatologist whose practice was in Mobile, had done his internal medicine residency at Keesler and volunteered to help with the evacuation. With some preparedness and combat casualty training in the Air Force, he knew that the most need would be "from illness and hygiene and disease rather than from the direct impact of the storm. I knew what to expect and how to deal with it."

Scroggie gathered supplies from his own stock and also helped others on the medical team to raid sample cases, to get IVs and bags of IV fluids. The practice manager at one site offered pallets of water and diapers.

Upon arrival at the school in Biloxi, he said, "We could smell the facilities had started to break down. Toilet facilities were not working at all. And it was extremely hot—the place smelled of sickness. We went room to room, classroom to classroom. A lot of the initial stuff was explaining who we were and what we had arranged for them, that we would evacuate from the shelter because the facility was not adequate, and we needed to get them to some place safe."

Many of the people did not want to leave, to go to Georgia. They did not want to leave Mississippi or their homes, even their temporary homes.

The medical team provided on-site care as they could. Scroggie said, "A lot of it was routine—gastroenteritis, managing patients with hypertension whose medicines were gone. We did not have insulin or other diabetes medication, but we could give them something similar to make sure they were controlled. We didn't see a lot of trauma—just some cuts and abrasions. Low-grade skin infections. Many of them were just people who needed to hear, 'You're going to be fine.' They needed us to be hands-on and tell them, 'Your lungs and heart sound good. You're going to be safe and OK.'

Chapter 21

Especially after the successful shelters' evacuation, Travnicek believed the state health officer and his underlings aimed to remove him from the County EOC and, possibly, even from his job as director of Coastal Plains Public Health District IX.

"Dr. Amy, working through Jim Craig, director of the MSDH emergency response, and a few other minions, was constantly trying to get me to leave my post at the Harrison County EOC and to allow them to assign some sort of newly minted medic-trained people to our vital desk in the EOC," Travnicek said. "Thank God they confined themselves to the WIC Center on Dedeaux Road . . . they affectionately called it the EOC-North. They had their very own center, talked to no one, and did nothing—and never came to the official and legal Operations Center, ever. Under the right circumstances, they could have been helpful—if they had the right direction and coordinated with the actual legal EOC in the Harrison County Courthouse."

Craig defended establishing the agency's command center for Coast operations on Dedeaux Road: "We kinda just *occupied* that space because it wasn't open anyway. People slept outside in their cars, on concrete slabs, and it was so hot—no power, no anything. The district office was not inhabitable. None of the county offices—we lost the Jackson County office; lost Hancock County; and Harrison County flooded. Dedeaux was as far south as we could get so far as a Health Department piece of property."

Nobody from the Central Office ever discussed the location of their command center with the physician legally responsible for public health

Elfer added, "Being in the State Guard—they do have a legitimate mission. Generally, it's to man armories when a unit goes on deployment, and they're also Red Cross shelter managers."

According to the Guard's website, the State Guard is "an all-volunteer organization tasked with supplementing the forces of the Mississippi Army National Guard and/or Air National Guard upon the order of the Governor of the State of Mississippi through the Adjutant General of the State of Mississippi. . . its primary mission is to assist in coping with any man-made or natural disaster" and is not subject to any federal authority.

"I didn't know that was their mission," Martin continued, "but two Biloxi police officers were there. The thing is the State Guard and the Mississippi Army National Guard wear the same uniforms. The only difference is the patch. And if you're not paying attention, you wouldn't know any different. So I told the Biloxi police officers, I said, 'Look. I don't think he's supposed to be carrying a weapon. He is not a Mississippi Army National Guard soldier. He's with Mississippi State Guard—that's out of our hands. . .'"

EOC Commander Spraggins got involved the next day, on September 4.

"General Spraggins said there were some problems at Biloxi High School involving the State Guard," Martin recalled. "And what we were told was that a doctor or someone there had been threatened."

Elfer said that "the FEMA response team that goes in to handle medical things got threatened by the guy."

"They had been threatened, and they came over to see General Spraggins," Martin added. "And they said they had been threatened. John and I already knew weapons were there, and this confirmed it. So we got a hold of the Biloxi police chief, and we met out at Biloxi High School with a SWAT team at 6:30 that night. We went into the school and disarmed—and that was funny. . ."

"It wasn't there for a minute," Elfer quickly reminded. "What you have to remember is that those SWAT guys had been up as long as we had and had been working the streets. The only reason they got called in was we knew there were weapons . . ."

And almost as quickly as the upset occurred, law enforcement intervened, resolved the situation, and went to the next assignment.

§

The separate command center for the State Department of Health, the very idea of EOC-North, displeased veteran hurricane responders. "That was stupid," declared Steve Delahousey. "Nothing calls for the State Department of Health to set up their own command center. They gave me and Dr. Travnicek and Hargrove nothing but trouble. They threatened Hargrove.

"No other state agency set up. It was so dysfunctional. And they gave those guys guns. When they walked into the EOC with side-arms, I asked, '*What* are you doing?!' We were close to anarchy. 'I don't want my nurses and doctors in the EOC to see nurses and doctors from the Department of Health in this courthouse with side-arms. You get the hell out of here, or I'll kick you out.'"

Hargrove also objected. "I walked into the courthouse, and two of them came with handguns on their sides, and I asked, 'What are you doing with handguns?' And they said, 'Well, we fear for our lives.' And I said, 'Well, I don't understand.' They said, 'We heard gun shots where we were set up.' And I asked, 'Where is that?' And they said, 'At the WIC Center on Dedeaux Road.' And I said, 'Well you're going to hear gunshots there on a normal day out there—that's just that area.'" Then he addressed Delahousey: "I cannot believe they've come down here wearing guns. They're the health department—what jurisdiction, what authority do they have to wear guns? They'd better hope that something doesn't happen or that they don't end up shooting somebody because that's going to cause a bigger problem. That will cause a riot. The health department is not a police agency—if you want to wear a gun, be a cop!"

Word reached Craig at his office in Jackson. "That would have been, probably, Art or Mills. I don't know why. I can tell you what I think I know, but I don't really know. There were a lot of rumors going on, more than anything else, about civil unrest, I think mostly in New Orleans, but it got played in Mississippi. . . When you don't have a good communication system, I think, unfortunately, that the rumor system is what's in place. Both Art and Mills are military; they are both authorized to carry firearms in the State of Mississippi and received same from the director of Public Safety. It's not a question of whether they had the legal parameters to carry the I-400 class and are absolutely adamant with that class that incident command is about command and control, and that occurs at the county level. Katrina was the first real test."

§

"A day or two later, after we closed that shelter, I received a call from Dr. Amy ordering me out of the EOC," Travnicek recalled. "Apparently Steve had realized that I could no longer be trusted to clear all bathroom breaks with him or otherwise do what I was told. In his mind, I had become dangerous. I learned later he had called the director of emergency operations for the health department and told him I had lost my mind—that I was 'stressed' and needed a break. I got a call from Dr. Amy telling me that Steve had said I was stressed and needed a break. *Not on your life.*

"Since I saw the local political officials every day, I let Connie Rocko, Larry Benefield, and other members of the Harrison County Board of Supervisors know what Steve had done. I think Connie and other power players in county politics back-doored the governor or his staff. She told me she also spent some time actually talking to Dr. Amy; I will always be grateful to her."

Rocko went to bat for Travnicek. "He was there—he never left," she said. "And whenever we needed him, he was there. He'd say, 'You can't do that. You cannot flush the toilets. . .' Everyone came to the Courthouse; that's where we needed all the vaccinations, and the people who worked there needed the vaccine for tetanus. We had first responders there who were going to be out in the rubble. I don't know what the heck happened to the guy in Jackson, but we couldn't get any vaccine; he wouldn't give us any. Then he tried to take Dr. Travnicek away from us. And I saw him and I said, 'You've got the best employee I've ever seen, and thank you so much for having him here.' I didn't know what all was going on."

Simultaneous events happened that day at the courthouse and at EOC-North, but Travnicek could not be in both places at the same time. With the order from Amy—and according to Craig, an "order" held important status in the agency's emergency response operations—Travnicek reluctantly left his station and drove north. "I did what I had to do to keep my job, at least temporarily; so I left immediately for the district office and waited for them for about two hours. Everyone in the department feared that he was coming to fire me. They left with heads down, not even a hello. They were scared to death of Dr. Amy and had told all the leadership in Jackson I was history. I waited, and I waited. At one o'clock, he arrived.

"The key to understanding the situation is this," he explained. "At the one o'clock press briefing, I normally sat next to General Spraggins to answer any medical or medically-related question—and there were a lot. I had intended making that meeting because rumors of dysentery had reached the public.

"I waited one hour, two hours, and finally in came Dr. Amy around one o'clock with Dr. McNeil. In a brief meeting, maybe 10 minutes, he asked how things were. He did not wait for an answer. He never waited for an answer. He went on to thank me for my service to the state, which he did to everyone. Then he told me I was to report to Dr. McNeil and do whatever he told me to do. Then he left. I think he thought Dr. McNeil was going to order me out of the EOC, or fire me, or both. As I said, Dr. McNeil was no fool, and he knew I was right. The whole conversation with Dr. Amy probably took less than a minute. He walked out. Dr. Amy would threaten your job on a daily basis and in the next sentence praise you for some ridiculous reason and then walk out of the room, which he did.

"I looked at McNeil for a minute. I don't recall what he said, certainly nothing useful. I said I was late to the one o'clock briefing and that the shelter epidemics would be an issue. I desperately needed to be at General Spraggins' side. I left and returned to the EOC."

With that, Travnicek regained confidence in himself and in his job. Although he and McNeil continued to cross paths, further conversations were short and, for Travnicek, without substance. "Quite frankly, he had done what I needed him to do; he gave me credibility and a vision of what was going to happen, and it saved the Coast from a disaster like New Orleans."

Spraggins always deferred to the physician for any media question related to medicine or public health. For the first four or five days, all went with clear information and correct answers. The briefing on this day was neither smooth nor totally true.

"Too late," Travnicek described his return to the EOC. "General Spraggins had responded to the media, including *The New York Times* and *Boston Globe*, when he was asked about diarrheal illness in the shelters and on the street. In desperation and not having me at his side, he said it was 'not serious.' He said it was like 'the old dysentery from the military service.' The only thing the press heard was 'dysentery,' and they went ballistic. I think Joe was mad that I was not there; even when I told him I had just survived a meeting with Dr. Amy and Dr. McNeil, he was not very happy."

Whether unhappy that day, General Spraggins later had only praise for his public health doc. "Dr. Amy strapped him. He was not given the capability to do what he needed to do. He would come to me and tell me, 'This is what we need to do, but I can't get it done. You're going to have to call Dr. Amy to get it done—they won't let me do it.' His hands were tied. And he wanted to do right. Amy stopped whatever he could. Travnicek is wonderful. He wanted to do the right thing. It was almost like Dr. Amy wanted to not turn over anything to the counties; he wanted to do everything himself. I'm not saying he was deliberately trying to hurt anything, but his management style was control. And it strapped us.

"As a matter of fact, just to give you an idea, there's a center on Dedeaux Road that takes care of babies—WIC (the Special Supplemental Nutrition Program for Women, Infants, and Children). We begged for days to get access. We begged to get in there and get help for babies, and they would not give us anything, so we finally broke in there to get what we needed for the babies. We did not have stuff for the babies, and we had to take it upon ourselves to go in there and get it. . . We could not get a damn thing out of state public health."

The EOC Commander also ran into resistance regarding tetanus vaccine. "We called," Spraggins said. "There was something in the neighborhood of 15,000 vials of tetanus vaccine in the whole state, and we told them we needed at least 50- to 100,000 vials in Harrison County alone, and they couldn't get it and didn't want to get it. So what did I do? I called Stan from Air Bus. I called him, and he asked me what else we needed, and I told him we needed tetanus vaccine, and the next day we got 50,000 vials of vaccine for tetanus coming here. And do you know, Dr. Amy intercepted them and wouldn't let us have them?! We had to fight to get those vials.

"I tell you what—that man! I don't talk bad about people, but he did not have the big picture," the commander declared. "He needed to have the big picture, and he didn't have it. I don't know what the problem was, but . . . I'm sure he meant well."

Another problem that surfaced in the EOC involved the father of one of the public information officers, a nephrologist whose patients desperately needed dialysis. "We did not have water. He needed water. We called state public health to get water for dialysis—they'd been days without dialysis. Public health wouldn't do it. They sent a damn bus down here and said to load them up and take them to Jackson. Half the people were not going to go; they weren't going to leave their loved ones and go. . .

"Their answer to emergency medical—they wanted us to take all the special needs people and go to the Navy base in Meridian, or somewhere in Meridian," Spraggins continued. "That was not going to work. The danger of travel. If my mother was in a wheel chair, do you think I'm going to send her up there without my being able to take care of her? The storm's over now. Instead of setting something up close by, they still wanted to move people out. And you can ask Dr. Travnicek. We still probably have a tremendous need for special needs that we still can't get done because of the state. Maybe we need some house cleaning up there! You know, those guys from up there showed up down here with pistols on their sides. They didn't have any authority! They were walking around here as if they were taking over the world, and they did not have the authority to do that."

Chapter 22

Harrison County EOC reported two fatalities in Gulfport as the "first known related to Katrina." One person was missing, according to press release 31 on August 29. For Coroner Gary Hargrove, locating storm victims and managing rescue and recovery was top priority. First elected coroner in Harrison County in 1996, he knew that people, particularly the news media, would want to know about search and rescue: how many dead; how many rescued; how many missing. Just one day after the storm, newspapers reported 80 dead and "damage to Coast in billions." Headlines proclaimed "flooding, wreckage, death."

Hargrove knew that he owned primary responsibility for communicating the truth; he just had to get organized, get some help, and get about rescuing and recovering every person possible. He had plans in place about where to set up DMORT (Disaster Mortuary Operational Response Teams), how to do things, and what he needed. "Those plans kinda went," he acknowledged the obvious. "Katrina took everything out. So we met with the general; he was able to give us a spot at the airport—we put it in the no-fly zone and started the operation the Tuesday after the storm. We got the operation going and started bringing people in and trying to get them identified as quickly as possible and return them to their families."

Some news media were already reporting thousands dead, and Hargrove aimed to stop the dissemination of incorrect information.

"I gave the numbers to General Spraggins the first day, and I told him, 'We're not going to speculate. If we do speculate, all we're doing is creating a bigger problem. Let's report the truth to the people—how many people we recover; if we rescue somebody, we put that out there.' And he agreed with me. Commissioner Phillips agreed with me; the governor agreed with me; the head of MEMA agreed with me. And that's what we did. We gave out true numbers of people we were recovering. After the first news conference of doing that, I was called back into the office and told, 'You will be here in the office in the mornings from now on at 10 o'clock; you will deal with this.' And I said, 'Fine.' So each evening, everybody would come in; we would get together and put the numbers together, get it all done, go back over it the next morning, put it in writing, give it to the general, and then go to the news conference. And we would notify the governor, Mr. Latham, and the commissioner of those numbers—and kept them totally informed."

Hargrove's systematic approach worked so well that in a meeting at the EOC that first week, Governor Barbour turned to him and said, "You're in charge." "In charge of what?" the coroner asked. "The southern six counties," the Governor replied. "Each county has its own coroner," Hargrove protested. No matter and no problem, the governor assured; he wanted one man in charge of the logistics. "You make it happen. We're going to send everything to you, and you get it out to the media."

His communication with other coroners and DMORT responders would be simple: "Report direct to us. Once you report to your EOC director, then you contact us."

After the first couple days, Hargrove realized he had not heard from Hancock and Jackson Counties. "The last thing we had heard from Hancock County was they were going under water, and they were supposed to have written their Social Security numbers on their arms. So we made our way over there to where they had the EOC set up. They didn't know where the coroner was, but I found her at one of the funeral homes there and told her what was coming, what was going to be available, and asked her what she needed immediately. We had brought body bags over to her, and then once we got a good idea of what she needed, we made our way back to Harrison County.

"Then DMORT came in, and we met with them. My first question to them was, 'Who's in charge?' And they specifically said, 'You are.' My second question, 'What *can* you do for me, and what *are* you going to do for me?' And then we went through and spelled out how we wanted everything to work. Our whole operation went totally smooth—every time, until the health department got involved. When the health department got involved, everything just slowed to a crawl."

Having encountered health department employees "with handguns on their sides" in the courthouse had stung the elected official. After questioning their authority and commenting to Delahousey, he returned to work, only to be notified via radio the next day that he was to attend a 10:30 meeting with Department of Health officials at their location on Dedeaux Road. Not much for meetings—"Meetings don't help me do my job"—he nonetheless appeared after making a couple recoveries in Biloxi.

"So I walked into a room where Jim Craig, Art Sharpe, Mark Chambers, and all of them were, and they basically said I would need to start reporting to them from here on out. 'We're ESF-8,' they said, 'and if any of the other coroners are not cooperating with you, we'll go in and take over and insert our own people.'

"You have to understand," Hargrove explained to them, "we coroners are elected officials. The only way we can be removed from office is to show malfeasance or to be on a recall. That has not happened, and after a disaster like this, you're not going to have that happen. No coroner is going to take over another coroner's jurisdiction."

Empowered and shouldered with multi-county responsibility by the governor, Hargrove worried because "the health department put out that I had been placed in charge over everybody, and that everybody would answer to me." Although he considered the obvious miscommunication a major problem, he decided to just get beyond what he saw clearly as a power play and to get everything up and running. One part of that "everything" encompassed missing persons.

"They wanted numbers on missing persons, and I told them no number was going to be right, that all the numbers are being duplicated," Hargrove said. He told them no one agency was collecting information about persons reported missing, that Red Cross, Salvation Army, the State, and three different family assistance centers in Hancock and Harrison Counties were all counting but not coordinating. "You've got 2,000 people missing here, we've got 5,000 missing here, and it's coming from a multitude of areas." Although he initially was comfortable using numbers from the counties as his "official missing persons" count, he knew the EOC command staff expected more.

"I walked over to the EOC," he remembered. "I never will forget. They wanted numbers on missing persons. And I told them, 'Everything is being duplicated. The numbers are not going to be right.' There was a gentleman from Florida who said, 'I can get it sorted it out, put it all into the computer, and then we can match them up and come up with one list.' I turned my information over to him, and I told him I didn't want it going to the media until we got something solid. And I walked out, and he turned it over to the news media immediately, which created a firestorm because there were people's names on it that I knew were alive, but I had not had time to take them off the list.

"Trying to deal with the news media," he just looked down and slowly shook his head. Some time later, about six weeks into disaster response, he would address discrepancies in the missing persons numbers through creating a Missing Persons Task Force.

§

MEMA Executive Director Robert Latham and his federal counterpart, Bill Carwile of FEMA, also went to Hancock County. On the ground in Gulfport the day after the storm hit, they had no communication with anyone in Hancock County, Pearl River County, or Jackson County.

"So we would write some questions down on a note card, put somebody in a car, and say, 'Okay, go to Hancock County, get answers to this, and bring it back to us.' That's how we got our information," Latham described. "On Wednesday, Bill and I were so frustrated that we didn't know what was going on. So we got in the car and onto I-10, drove on, and came off onto 603, which is the road that leads from Kiln down to the Bay of Saint Louis. Once we got off 603, we had mud and stuff this deep (indicating two feet); we dodged power lines and houses.

"We finally got to the Hancock County EOC, which we knew had been destroyed, and they actually had a little push-up tent out front that they were operating out of. Bolivar County had gotten there with Delta State—they had a bus—and set up beside the tent.

"So I get there, drive up, and Congressman Gene Taylor was there. And, you know, his house was totally destroyed. He comes out, and he looked like a homeless person; well, he was because he'd lost everything. But he'd been out trying to help people, and he came out and met with the EMA Director Hooty Adam, some of the Board members, and Chancery Clerk Tim Keller, who was also the county administrator at the time. Congressman Taylor came out and said, 'Robert, I've got somebody I need you to talk to."

Inside the bus, outfitted with tables and chairs, Congressman Taylor introduced Latham to a man in tears. He was the local funeral home owner, he told them. "My morgue is full. I don't have any more room. I'm going to have to start putting bodies down on the sidewalk and cover them up."

Latham knew that was unacceptable. Returning to the tent, he asked about refrigerated trucks that had been sent toward the coast before the storm's landfall. He knew that on the previous Saturday, Carwile's team had ordered eight reefer trucks sent to Mississippi.

"Not only had we not gotten the reefer trailers," Latham lamented, "we hadn't gotten anything else, either. Stuff just wasn't there. Bill said, 'We don't know where they are; we requested them; they're somewhere in the pipeline. What about all the reefer trailers that were at the Port?' They were all piled up in the middle of Highway 90—not a single one over there was usable."

Serendipitously, Latham turned to see what he remembers as a nice, shiny stainless steel truck as it arrived. "I never will forget it. The driver walked up, asking for directions. I said, 'Is that your truck out there? What are you hauling?' He said, 'Well, I just dropped a load off at Stennis.' After Katrina, that's where we set up our staging area. He was basically a contractor hauling ice; so I asked, 'Where are you going?' 'I'm going to get another load of ice and come back.' I said, 'Can I rent your trailer?' 'What do you want to do with it?' And I said, 'We need a morgue.' He said, 'I would love to, but this is the way I make my living, and I won't be able to do anything but that again if you use this trailer as a morgue.' So I looked at Bill and asked Bill, 'Can we buy this trailer?' He said, 'Absolutely. We'll figure out a way.' So we sat there with the county administrator and negotiated a price of $25,000; we bought the reefer trailer—I co-signed for it—and that's what we used as a temporary refrigerated morgue."

Latham uses that story to illustrate the range of responsibility and opportunities emergency management officials encounter. "Don't ever think that's something you won't have to do. Or ever think that's somebody else's job. I don't mean just morgues—I mean anything. I never thought in my career I would have to figure out how I was going to store dead bodies, but at that point, we had to do something, and we were fortunate the guy drove up. . . And they've still got the trailer today in Hancock."

§

Initial reports of more than a thousand deaths showed that "reporters were not listening," Coroner Hargrove said. "They looked at the devastation of the homes and said, 'Oh my god, the whole coastline is wiped out. You've got poor families; we figure this many houses per block, and this many blocks long.' . . . I think that's where they came up with their numbers—unless they were just pulling the numbers out of the wind, and I would hope they wouldn't do that. I don't think it was done to hamper what we were doing, but one thing I do know is that at one news conference, the general did his part, and when I got up to do my part, I gave the numbers out, and WLOX asked a question. I turned and looked at them, and I answered the question. The *Sun Herald* asked a question; then WLOX asked a question. I was still looking in that direction, and a young girl from the national media was sitting off to my left in the audience. As soon as I got done, she said, 'Don't you think it's true that you're not reporting the correct numbers, that you're covering up?'"

Deputy Coroner Misty Velasquez, a veteran in working with the news media, gulped from the back of the room. "Gary was mobbed by the media," she said. "I sometimes would actually throw my body between him and the media because they were really aggressive."

Hargrove still bristles at the thought of that reporter from New York: "It was like somebody put a laser on her because I turned and looked at her, eye to eye. Even though I was looking directly at the girl, Misty was in the back and I saw her reaction. She later told me, 'The girl from MEMA said, "Watch it. He's fixing to unload on her." But I didn't. I looked at her and said, 'I am really sorry'—and I had no clue this was going national—'that you and your agency can't deal with the truth, and that you would prefer that the numbers be really high on the death toll, but, no, we as leaders are really glad that it's not as high as you want it to be.' Total dead silence. Then I said, 'Thank y'all very much. I'll see you tomorrow.' And I turned around; there was a guy from Associated Press—he always sat off to my left—and he always called me Gary. I had never met the guy personally, but he said, 'Gary, thank you very much for that.' I just kinda looked at him. Misty stopped me and told me I needed to stay stopped because when I walked out I was fixing to get hammered by the media. I said, 'That's OK; I'm over it now.' She said, 'Well, let's try it.' And somebody with the *Sun Herald* tried to talk; I said, 'If you're going to talk, you're going to have to walk because I've got work to do.' She asked me, 'Do you expect the count to be in the hundreds or thousands?' and I said, 'No, I do not. Based on what we've got going so far, I expect it to get to maybe 100 or possibly 200, but I'm not going to speculate. We're going to go with what we've got: the facts, the true numbers.' We took them from search and rescue to search and recovery to monitoring debris and recovery; I kept them informed."

§

Hargrove set up his command center in a boat hauler—"the same thing NASCAR has, but for boat racing"—that his buddy Steve Page drove in from Fort Myers, Florida. Filled with food, water, drinks, clothing, and baby stuff and set up across the street from the courthouse, the rig and Page stayed until November. Volunteers would go out during the day with a truck or trailer loaded with food, water, and other supplies and drive through the neighborhoods to give people what they wanted. "He tried to donate it to Red Cross, but they wouldn't take it—said they wanted money. Which kind of left a bad taste—this group in here now is trying to make amends, but when you're trying to do something to help your community, and they're fighting you all the way to Washington. I was already on the outs with them because they wouldn't even give us access to who was in charge; we did finally find out, but it took an act of Congress to do it. They fought us on the local level; they fought us on the state level; and they fought us at the national level. We took it to Washington to try to force them, and they kept saying Privacy Act, HIPAA, blah-blah-blah . . . Finally, I just told them, 'The next time somebody tells me that, we're going to have a problem. You're interfering with what I'm trying to do, and that would be interfering with a death investigation—*sorry!*'"

For all the individuals who did not understand the scope of his responsibilities, Hargrove knew he could count on Travnicek to help.

"When we started getting people in here, I called Dr. Travnicek and said, 'Hey, these guys are tromping through God only knows what; we're going to be dealing with bodies; and I need to make sure they're inoculated, to protect them.' Dr. Travnicek was probably the biggest help to me in trying to not curtail but to help. I kept him informed because he's in that loop of our EOC and if I needed advice about handling human remains, I went to him. That's what he's for: he's our health officer, and he carries a great knowledge of diseases and other stuff after disasters. I trust him. Obviously, I wasn't getting anything out of the state, which was a major problem for me simply because it was more about 'let me pat myself on the back and be the big guy.' I wasn't impressed with the health department; I was not impressed with the people who were down here. They were having food catered to them from Jackson; they were up at the WIC Center and calling it the EOC. Well, there is only one EOC in the county, and it's county-run and county-owned. You have an emergency management director who's the top dog and surrounded by the people there to help him.

"And here they were, running around doing their own little mission or whatever they want to call it, and not coordinating anything. We were having people come here, saying, 'Can you tell me where the WIC Center is? I need to get my credentials.' 'What do you mean the EOC? The EOC is across the street.' 'Well, we were told it's on some road named Dedeaux or something?' 'No, the EOC's right here; and if you need to report in, you need to go here.'"

$

A war zone—that's what the Coast became in the eyes of many, particularly Deputy Coroner Velasquez. "And it's intimidating," she confided. "The EOC had become protected by the National Guard, and we had all these military people and no lights—except the extremely bright lights the national networks brought in with their trucks."

But the national media soon moved on, moved west to cover the flooded City of New Orleans. "Rescue lasted only a few hours," Velasquez realized.

"Recovery started immediately. We were looking for both alive and dead bodies. On Howard Avenue that first night, it was one o'clock in the morning, and everybody was scared. One of the two-story houses had people who had stayed in the second story—that second story floated down the road, and they stayed there. Only when they spotted Gary did they come out. They didn't want to leave. They were so scared.

"Beyond anything that you can feel or create, it's the people in the community who make it special. We'd go to neighborhoods that we wouldn't know. We were south of the railroad tracks, in one of the old East Biloxi neighborhoods with Asian, African American, and Caucasian people all on the same street, and this one woman came out; she could tell us everything we wanted to know about everything and everybody and where they were—and about their kids and pets, too."

Alabama responder Sherry Lea Bloodworth would also go to East Biloxi, going door-to-door with physicians on her team to pull people from damaged houses. "They had dead bodies and standing water in their houses. People were crammed together in such small spaces and had no water—there was no water, no bathroom, no food, and bodies everywhere—the fourth or fifth day and nobody had picked up their bodies. I went back and reported that."

§

Several weeks later, after Gulfport visits by Hurricanes Rita and Wilma, the airport building was no longer safe; so Hargrove moved the DMORT project to the Gulf Islands Water Park site, city-owned property between I-10 and Landon Road. Urban Search And Rescue (USAR) originally worked from there but had moved to New Orleans, allowing him to work there with such remaining entities as the US Marshals, MBI, and Mississippi Highway Patrol. He also set up the Missing Persons Task Force at the new location. Broadly, the Task Force included not only his own office as coroner but also DMORT, Mississippi Bureau of Investigation, Georgia Bureau of Investigation, Kansas Bureau of Investigation, ATF, ABC, Bureau of Narcotics, the Federal Marshals, the FBI, Crime Lab. "Just a multitude of agencies came together under one heading that came as a brainstorm of Colonel Mike Berthay of the Highway Patrol and myself," he said.

identified, and returned to their families so they could shut the operation down by December 15. In charge and in command, Hargrove firmly told FEMA that once the operation reached that point, he wanted them to leave. As work progressed, he moved the deadline, usually by a day, and finally to December 10. With the date change announced, Hargrove was called to yet another meeting.

"What do you mean changing the date and not telling us?"

Hargrove answered clearly: "'I don't have to tell you—you're not in my structure of people I report to. There's General Spraggins, there's Colonel Berthay, there's the commissioner, and there's the governor. I'm sorry, but where do y'all fit in?' I said, 'I'm not reporting to y'all. You can get your information like everybody else is getting it. You want to know the numbers—the number of dead, the number of missing? Come to the news conference; watch it on TV. I'm not taking my time to track y'all down to tell you anything.' That didn't go over well, as usual. But I did what I had to do. And that just wasn't going to happen. I know my job, and the people elected me to do this job. If the people didn't believe in me, they wouldn't have elected me to office. And I wasn't going to let somebody who didn't have a clue about what we do and about what my job is come in here and tell me what to do. My training far exceeds theirs."

Chapter 23

Misery spread from the Gulf Coast throughout most of Mississippi that first week after Katrina's landfall in Hancock and Harrison Counties. The September 3 edition of *The Clarion-Ledger* proclaimed "no end to desperation." The newspaper focused on unacceptable results, grief, pain; the "gone" homes, cars, jobs; and hungry, thirsty refugees. Reporters wrote about the massive effort to provide food, water, and sanitation.

A story about the poultry industry spotlighted the urgent need to dispose of millions of dead chickens.

Obviously weary and under intense political pressure, Governor Haley Barbour promised, "We are going to rebuild bigger and better than ever. It's going to take some time, and people have to be patient." He described "utter destruction far beyond that of Camille, a sea of debris from Waveland to Pascagoula on the beach and inland for several blocks to the railroad tracks." Cargo containers at the Port of Gulfport were strewn like matches, and most buildings were "literally wiped away."

Sunday's *Clarion-Ledger* listed the statewide death toll at 167. An article on page 5A quoted District Health Officer Bob Travnicek: "This is not a hurricane. This is not a disaster. This is a catastrophe!" He talked about tainted water having been suspect in a special needs shelter's sickness outbreak and assured that the shelter had been closed Saturday. One headline declared the Coast "battered beyond belief," and yet another article mentioned the "strong stench" 30 Urban Search And Rescue (USAR) teams experienced as they looked in lumber piles, roofing material, and concrete for bodies and parts of bodies.

The public health doctor covered public health concerns. "Survivors need to pay close attention to personal hygiene, stay away from floodwaters, and avoid dead animals. We have to look seriously on how we're going to make it day-to-day. And we're also concerned about carbon monoxide poisoning from diesel-powered generators and for chainsaw injuries by people clearing storm debris."

For that September 4 edition, Governor Barbour offered his opinion: "From the Civil War and reconstruction through the Great Flood of 1927 to Camille, Mississippians have a long history of coming back from disaster. Each time our state has come back stronger than from the previous episode. We're tough, resilient, caring, loving people. We're literally the most generous and charitable of Americans, but this time our generosity is being matched with great generosity to us from others. The federal government, our sister states, and private citizens and companies are overwhelming us with their help. And that's critical because it will take a joint effort, a magnificent collaboration to rebuild.

"And rebuild we will. We will rebuild the Coast bigger and better. We will rebuild a Mississippi that exceeds anything we've ever known—a state with more prosperity, more opportunity, and more equity for more people than ever in our history."

§

Deputy Sheriff Rupert Lacy-turned Logistics Chief Rupert Lacy found humor in seeing himself and Harrison County colleagues step into new boots after Katrina.

"The larger the event, the more likely you might get thrown out of your element. You may take on some type new responsibility. I laughed as we set up an Incident Command Structure," said Lacy, one of only a few locals who actually had taken NIMS-ICS training before Katrina. "Florida said we had to do this. And we knew we had to do it. We started plugging people in. They couldn't even spell ICS, but they started working it, and many city and county employees took the ICS course or some of the training online as they were working the event!

"The stamina of our residents—we had some people, city employees, who were assigned to the group, to the big picture. They didn't have an office to work in for the city. So they showed up here to come to work. Some of these people didn't even have a roof! They didn't have a home to live in, but they showed up to work. They showed up on their normal job and were reassigned to work here. That first week, I know I was going until between one and two o'clock in the morning. I'd lie down in my Tahoe, my Sheriff's Department Tahoe that had very few glass windows left in it. And I'd recline the seat back. Everybody laughed about it because it sat out here in the west parking lot, right next to the portalet. I couldn't tell you if a portalet stunk; everything smelled the same to me. But for two to two-and-a-half hours, there was plenty of time for me to rest. If I could take my boots off, if I could relax—my radios were in here being charged, and if something major happened, the people in here working knew to send the sheriff's department security people out there to get me up."

Lacy looked for some sense of "normal" but knew that would be long coming. In the meantime, he settled for "routine." The days started at 5:30 am with a trip to the men's bathroom. Inmate trustees had ridden through the storm with him and other first responders in the EOC. "One of them would meet me in that hallway and tell me, 'I was able to get you another bottle of water.' Therefore, within a week, maybe longer, I could shave, shower, and brush my teeth with 16 ounces of water. And if I got extra water—man! That was the greatest thing in the world! I'll never forget—we had some water come in that was in the five-gallon jugs. There were about four of us. We knew exactly our time in there. We had one shower stall. And he said, 'There's a five-gallon bottle in there, up on the ledge. You get one-half that bottle.' I was able to wash my hair, to put soap on my whole body! I laughed about it. And Gillette shaving cream is in my go kit now. I like Gillette with the flat round cap. I'd fill it up with water, wet my face, and I'd put my shaving cream in there, and I could wash my razor and shave my whole face with that, in that little cap. . . I thought about growing a beard, but I can't stand that first week. I had to shave. I had that bottled water for brushing my teeth.

"By day two, we were happy because we have a well here at the building— it's not a big well and, of course, the sewer lines were down. But we quickly found that our little well put in in 1977—it was great then, but instead of a 100-gallon source tank, we need 500 gallons now, and it needs to be gravity fed; we know these issues now. But that was a perk. Having water—that was a plus."

Lacy, ever the perfectly pressed and dressed, dismissed anybody's caring about clothing. "You didn't care about what you wore, whether it matched. But I knew to change socks. If I wore a pair of socks today and knew I had X number left, I'd take them out there and find a place for them to hit sunlight and to air out. I laid my socks out in the back of my truck. I couldn't drive it anyway; it had no window in the back and three flat tires. But the wind could blow through and . . . Everybody laughed at me."

§

Bob Travnicek had his idiosyncrasy, too. He remembers that every day for the 100-plus days he worked inside the bunker he wore the same blue golf shirt, a shirt emblazoned with Harrison County Emergency Management Agency's logo, and on a custom lanyard his Mississippi State Department of Health identification badge with its iconic faces-of-Mississippi logo. To him, that shirt—which he still wears for special occasions that commemorate Katrina—embodies all the struggles, the strife, the success of the Gulf Coast warriors against the monster hurricane. To him, that shirt contains the secrets he harbored as he labored against all odds to do the right thing, to save lives, to prevent social unrest.

Whether he had escaped a few hours to sleep at home in Long Beach or had curled up with a blanket and a pillow to sleep under his desk in the EOC—"And he snores like a freight train, by the way," AMR's Doyle laughed—Travnicek paid particular attention to the mood inside the EOC. "FEMA Mike is incredibly gifted," he said. "And not unlike Billy Crystal—he's naturally funny. But he can laugh or chew your ass. He and Spraggins threatened not to be incompetent and to keep people from killing each other. And that's a major deal. The road guy was also very funny, like Rupert Lacy and Bobby Weaver. But when the crunch was on, we had to kick ass and take names. This allowed us to cope. Sometimes when everything would get really intense, I'd stand on my chair, and I'd ask, 'Is everybody having fun yet?' I do things on purpose. I measure the mood of the crowd and then go do something stupid to make other people feel more comfortable."

On a particularly memorable morning—a Sunday morning filled with a general sense of his own drowning in minutiae—Travnicek hurriedly turned to answer an incessantly-ringing telephone at the public health desk.

"Some guy called me up. I don't remember," he admitted. "I don't remember your name when you walk out the door, because with me, you live in a moment; you're dead before you get down to the streets. I'm sure: the future is uncertain. I don't care about the future, so this is it. So, the guy called me up and he said, 'Well, I'm from the CDC,' and I said, 'I don't care if you're from. . . I don't know why you're calling me.' 'Well, we have some resources.' So I said, 'Okay, come on down.' 'Well,' he said, 'that's not what we do.' And I said, 'Well, what do you do?' And he said, 'Well, we can give you computers.'"

That, to a physician who would have been grateful for two sharpened number two pencils and a new notebook.

"'Why would I want computers—what the hell do you think is going on down here?' I asked. I'm pretty plain spoken; I haven't got time for this. . . We don't talk a long time, but when you're done—so, at any rate, he said, 'Well, don't you need computers?' I said, 'What the hell for?' And, 'Well, we heard you lost computers.' I said, 'Are you crazy? We were blown to hell and back. We lost health departments! The health departments are gone; they're slabs, the roofs are ripped off, and they're in Toledo! Where are you?' 'Well, I'm from Atlanta,' he answered, and I said, 'You must not have newspapers. We've been destroyed!' Then I said, 'I've got 25 calls; you're wasting my time because I don't want more angora sweaters.'"

Angora sweaters comprised a significant part of clothing donations delivered from all over the country to help Katrina victims. Even though appreciative for all assistance, first responders who knew about that particular gift questioned the wisdom of the donor—extremely warm garments for hurricane victims in the Deep South where average September temperatures hover around 90 degrees? Travnicek considered the offered gift of computers to be similarly irrelevant, maintaining his belief that "people give you what they want to give you, whether you need it or not."

Fortunately for Travnicek and the Gulf Coast, the charitable Charles Stokes did not hang up, even though he noted then and would recall that Travnicek "was a little brusque."

Stokes was president and executive officer of CDC Foundation, congressionally-established as an independent, nonprofit organization to connect the Centers for Disease Control and Prevention (CDC) with private-sector groups and individuals. Stokes himself had taken a call from then-CDC Director Dr. Julie Gerberding, who suggested they activate the foundation's Emergency Response Fund to support CDC teams deployed to flooded communities and evacuee shelters on the Gulf Coast. She also asked the foundation to expand the fund's scope to provide resources directly to state and local public health agencies in the region.

Travnicek learned that Stokes had talked with Mississippi's former state health officer, Dr. Ed Thompson, who then was chief of public health practice and deputy director for public health services at CDC. Thompson said, "If you really want to know what's needed, call Dr. Bob Travnicek." And on that slow Sunday morning, with only a skeleton crew in the EOC in Gulfport, Stokes had prodded the local health officer. Stokes would remember Travnicek's saying, "We're dealing with dead bodies in the streets here, and I seriously don't know what we need yet."

Several days later, a surprised Stokes got a call back from Travnicek: "You know, what I really need is a building. In fact, two buildings for Hancock and Jackson Counties." Somewhat taken aback by the size of the request, the foundation's progressive director promised to think about it and let him know. But Travnicek firmly instructed that Stokes never again call him or even admit to having spoken with him. "Never mention my name," he said, "and never, never, ever, never mention Dr. Thompson's name because Amy will allow no help. Call Kathy Beam."

Providing contact information for the district's administrator, Travnicek dismissed the call and the caller, turning his attention to issues on what he considered his front burner.

Chapter 24

News reports that Labor Day—September 5, 2005—focused on both successful saves and human deaths. In Gulfport, the rescue of two Vietnamese men who had been tossed ashore inside their boat, 200 yards inland and at a tree line, and survived a week on hot peppers and water. They were diagnosed with classic post-disaster depression and many infected wounds. For responders, the new first priority was to discover and identify the dead, a job that could take up to a year.

Coroner Gary Hargrove was in charge of that for not only Harrison but all six coastal counties. Usually a solo practitioner, he had named the county's tourism marketing director, Misty Velasquez, and good friend and Orange Beach fireman Sam Jackson to be his deputies. His administrative assistant, Joy Yates, was second in command. Yates handled and kept him apprised of all routine, day-to-day operations, while Velasquez took responsibility for helping him coordinate disaster response activities. Of the young woman thrown into a totally different world of work, he said, "She did a really awesome job. Wherever I went, she was right there. If I tromped through standing water in gasoline and oil, she was right there; if I climbed over barbed wire fences, she was there; if I climbed over remnants of houses washed off the foundation and not knowing what was under there, she was right there. She was a big help—not only to help get the job done but also in coordinating information coming in, getting things ordered in time so they would be here the next day, keeping tabs on me on the disaster side."

Velasquez had what she calls her "hit the wall" day on Labor Day, her mother's birthday. With little-to-no sleep, food, or accustomed comforts, she had been "pressed into a world I was determined to do right by. I didn't have any background in it, and I had to figure it out." With 976 square miles to cover in Harrison County alone, Hargrove busy dealing with policy, and knowing they lacked resources, she "was getting everything from, 'Misty, we need to get our toilets cleaned out' to 'Hey, how are you going to get comments on this?' It was all on me, and I didn't have anybody here at the time who could help, to delegate it to. And then the Navy told me my boyfriend at the time was missing. . . Doing without resources and having to fight for everything, I was just overwhelmed; so I ended up with my head in the toilet for a little while, and then I got up and headed back out."

An Arkansas native who moved to the Coast in the mid-1990s, she called home as soon after the storm as she could get a signal on her cell phone. "Of course, my mom went into full mother mode and said, 'I'm coming down,'" but she was able to stop that. Conditions were too awful. "There are no words. I'm an excellent communicator, but there are no words to describe the Coast that day. There's no movie strong enough. There is no way the average person can understand." A California friend who visited a year later wondered why nothing had been done toward cleaning and restoring the community. Velasquez showed a destroyed port, described washed-out cemeteries, and explained that much money already had been allocated and spent to restore electricity and provide clean water.

"Everybody thinks we're just fine," she said, "but the national news coverage fell off, and mostly people think Katrina happened to New Orleans. I don't think people will ever—except for people who were in the Depression or in New York City for 9/11, and even then—understand the hope, the humbling experience of local officials going to a grocery store that had been completely destroyed and waiting for somebody to take a sack of potatoes so they can cut them up and fry them to serve to people standing in line because they have no food at home. They have no home, and they're standing in the street that's destroyed in front of a building that's almost non-existent just for a little bit of potatoes. I don't think people can grasp that."

The first 48 hours after Katrina's landing turned Gulfport into "a community that's taking care of itself," she said. "Let me tell you: federal and state were not here!"

When outside help did arrive, even Urban Search And Rescue (USAR) could not give precise timelines for their work. "'You have the distinction now of just having had the nation's worst disaster,' they told me. It was a challenge like no other, and they said there is no plan for this—'The plan we're working off was developed for the rest of the country, and it does not fit this storm.' They were just being honest.

"It's one thing to be objective and another thing to not be emotional when you're picking up somebody's mother," she described on-site experiences. "You have to put yourself in a mode that this is the task, and this is what has to be done. It gets easier, of course, but you always think 'what if?' One day we were out at the Mission Church on Howard Avenue, and we stayed there eight hours and processed 11 to 16 bodies—here's one, here's one, here's one . . . but the debris pile is so high, and we had to get cars moved."

Responders eventually recovered 97 bodies in Harrison County—all within two miles of the coastline. "The debris is what gives me nightmares," shuddered Velasquez, who worked most closely with both National Guard and USAR responders. "Debris was a large process that took many skills. As soon as we could get a breath, one of the USAR guys and I sat down in a room and created a debris body program, which we were able to get some resources for." They detailed where the trucks would be used and worked with contractors and spotters to recover every body possible. "Debris has a totally different definition for me—knowing I've got to walk and crawl up it and walk around and fall on it, knowing that debris really is everything or the equivalent of every effort by a person in their life. And it's their homes, their cars, possibly all the things they cared about."

Velasquez and Hargrove cared about the debris, but they had a bigger, overarching goal from before Katrina's landfall. "No family was going to be left with anybody missing," she affirmed. "We were going to get everybody back to their family. I guess I got a little possessed for a while; I made us and USAR go back into some neighborhoods several times if I thought we might have not done it right."

EOC Deputy Commander Linda Atterbury applauded Hargrove and every individual who worked with him. "It's very emotionally demanding on the human side of you, but the physical demands on most people—my God! To try to get into structures and check through them, the massive rubble, the dangers to themselves . . . the working conditions were horrible; it was oppressively hot. Of course, that's what you expect after a hurricane. They had no real relief at night when they got back to the courthouse because we didn't have air conditioning; we bathed out of bottled water for weeks."

Two of then-Congressman Gene Taylor's staff also praised the coroner and his work. "Gary Hargrove had a big job, a really, really tough job," said Bill Felder. "I'd be over there and see him coming back for lunch; these guys —you could see their faces, and I don't know how they did it. You could tell when they'd found several that morning because they'd come walking in and get something to eat and go over by themselves. And as it went on, and it got to a point where they weren't finding anybody, their demeanor was different. They started talking to each other. But you could tell in the beginning every time they came in that they'd found more folks."

Beau Gex added, "Search and rescue was absolutely phenomenal. I admire the hell out of those people. It would really strike you when you would drive past the courthouse and see that Dole banana truck and see them bringing bodies in. Seeing the Dole trucks used as reefer units, seeing their trucks being used as a morgue, and seeing the work of Gary Hargrove and the death dogs, and seeing them leave every day and knowing that had to be the worst job that anybody could do—absolutely phenomenal! To this day, I tip my hat off to them, and to Gary Hargrove; he is superb."

§

Search and rescue plus disaster mortuary operations comprised only part of Hargrove's responsibilities that year. Juggling day-to-day routine work continued, as well, and he also had to answer questions about missing persons. Beyond, Katrina unearthed and dis-interred Gulf Coast citizens.

"I had eight cemeteries that were damaged," Hargrove acknowledged, saying the greatest damage occurred at Southern Memorial. "I had five caskets that were displaced. And that jumped to 12 the next day, and the next day it jumped to 25. By that afternoon, it was at 50. I was being told, 'We've got it taken care of—don't worry about it.' But when it hit 50, I went down and saw the damage. I think overall it ended up being 232. We had water seepage on the beach, and the front of that cemetery was totally damaged. We had St. Paul over in Pass Christian; five houses washed on top of it. We had Live Oak that had heavy damage . . . We had displaced caskets, displaced vaults with caskets in them. . . I thank God that the Georgia Bureau of Investigations, DeKalb County Fire Department, and Birmingham Heavy Rescue came in initially and helped us—and the coroners and deputy coroners from across the state. They came in to help me get the job done."

§

Disturbed by missing persons stories she had heard since first moving to the Coast, Velasquez listened more closely to every fearful mention of lost souls. "All people talked about, every story you'd hear about Camille involved somebody they lost during Camille," she said. "They told me about people they were never able to identify because they were left on the island. But Gary and I had one goal: no family was going to be left with anybody missing; we were going to get everybody back to their family."

That commitment led her to study maps and convince the Coast Guard to give her a ride to Horn Island, the largest of Mississippi's Barrier Islands, which lies south of Jackson County.

With six Ohio firemen, she visually split the island in half so that no one would have to walk more than six to eight miles, based on the maps. Each team would survey the beaches, lagoons, marshes, and ponds that could make crossing difficult. About seven miles and two-and-a-half hours later, her cell phone rang. "Oh, I forgot to mention that there might be a body on the south side of the island," the caller said. "Oh! So we didn't need to come out here, to get to the other side of the island?" she asked. "Well, we did see a decorated island, scattered about with Tupperware and all kinds of little things—any little thing that washed away was on Horn Island. The casinos had these big pieces that had landed out there; it looked like *Planet of the Apes*. And more pelicans! All standing on one end of the island facing in the same direction. Talk about a beautiful view out there!"

And then they saw the casket, with feet sticking out. Several conversations with Hargrove and the Coast Guard later, she and the firemen managed to move the water-logged and sand-filled box into a boat and back to Jackson County, which had jurisdiction.

For most of the work, she said, "There's no time for slow or going back to check with the officials. We did have a few who came in from the state who promised to get supplies—'Tell us what you need'—and then disappeared and quit taking our phone calls. Then they would pop up because they wanted to be the one to tell the governor. You would be *amazed* how many people told me, 'I'm the go-to for the governor on this.' And they would power-struggle over who got this phone call first. . .

"The Department of Health stood out as an agency doing this. It was amazing to us that they seemed to end up on the top of the stack. The one who irritated me the most was Art Sharpe. He is uniquely interesting. They didn't know; they weren't around! It got to where they were trying to take over the authority of an elected official. Like, 'I just want to be in charge; I want my name on it.' It wasn't because they could bring more resources to make it better for us; it was all about who was going to get credit.

"Dr. Travnicek was very calm, which was very appreciated," Velasquez contrasted. "We had people from the outside coming in and screaming and banging on the doors; we didn't need such people in the EOC. Gary would go to Dr. Travnicek and say, 'I think this is what we need,' and he would help us get the resources. But he did not get the support he needed. He never said that out loud, but I had a hunch that maybe he had some of that going on. The people you put in the EOC are only as effective as their agencies allow them to be."

Chapter 25

Memorial Hospital of Gulfport executives reflected after Hurricane Katrina on the new meaning of "gone":

Gary: What was bad was when people would ask what it's like, and you'd just say, "Gone." It's like they never understood the word "gone." They were like, "What do you mean, gone?" And I'm like, Well, you know, you go to Long Beach, Mississippi, where the bank used to be with its concrete vault? Well, the cement vault is there, but there's nothing else there. It's just gone. "Well, what's next to it?" Well, nothing; it's gone. "Well, how's Long Beach?" It's gone. "What do you mean?" It's gone. It's not there.

Jennifer: I know, when I told my dad, he said, "Well, did you see my house gone?" No, Daddy, I couldn't get in there, but there's nothing out there. "What about the water? Is it there?" Well, we can see the water from the railroad tracks, but we can't get south of the railroad tracks because of that Concertina wire along the tracks. He asked me about everything that was in there for him. I said, it's gone, Daddy. "Well, what about the bank?" Gone; I can't say it any other way. It's just gone.

Larry: Dr. Roy Alston, the guy who's the head of the DMAT team, said it was the worst he had ever seen. Of all the hurricanes, disasters . . . he said it was the worst he had ever seen when they pulled in here.

Gary: And I still remember. Wednesday afternoon, the security guys took react to what comes your way."

As days and weeks passed, The Village expanded to also offer a laundromat with tables and chairs for playing cards and visiting, a pharmacy, and a book store furnished with donations auxiliary members sorted and displayed. Kids' clothes, cleaning supplies, boxes of donations appeared—an IT rep based in Atlanta and assigned to Memorial bought $10,000 worth of commodities and trucked it in two tractor trailers for the new community.

"The simple and short version of The Village," Memorial Hospital Vice President, Larry Henderson explained, "is that we had patients coming to the emergency room. The typical process is to take care of a little scrape here and another scrape there and, 'Well, okay, we've bandaged you all up; you can go home now.' 'Well, I ain't got no home.' 'What about your next of kin?' 'I don't know and. . .' We couldn't throw them out on the street. So we just started taking care of them. The long and short of it is that we were doing what needed to be done."

Beyond fulfilling their original mission as a hospital, Memorial did everything possible, said Diane Gallagher, "to make it normal in here because it was anything but normal out there! We encouraged our people to make this your stress-free place. We brought the SBA in here; we brought the Red Cross, FEMA, the passports people—you name it. We talked to the right people to get them on site so our employees didn't have to go struggle through all that."

Marchand added, "Honestly, the last thing we needed to do was pull a clinician or physician away from here to go arrange for a blue tarp to be put on their roof." Eventually, after surviving the scarcity of fuel for generators and getting power restored but before retail businesses returned, the hospital even set up a fuel supply station for employees. Despite that success, Marchand continues to choose not to reveal his fuel source for the time nobody else on the Coast could get fuel.

"When the DMAT team showed up," Henderson recalled, "they said, 'We can help you in one of two very specific ways: we can provide immediate relief to your emergency room physicians so they can take a break, or we can set up as an adjunct to your emergency room—set up outside and decompress and take some for the more moderate issues.' We took the second option."

Even though the staff saw a 400 percent increase in ER visits, the hospital never filled, and Marchand never declared a hospital state of emergency. By the end of 2005, Memorial had lost almost 300 employees, including about two dozen medical staff. In 2007, the management team continued to deal with availability of medical staff—they lost everything, and they left.

Constant helicopter traffic plus the presence of National Guardsmen, Coast Guard, and Seabees gave every sound and appearance of a war zone.

Jennifer Dumal was alone in the command center when she answered a phone call. "A really nice man asked if they could land their helicopter here, and I said sure. He said, 'M'am, it's a large helicopter.' And I asked how big? 'Like, really big.' Then he described to me how large it was, and I said, 'No. We're going to have to think of a Plan B.' But I didn't have any idea how big one of those was or what it looked like. Then we actually had three here at one time."

Marchand counted: "The Army Chinook in one parking lot, the Navy Chinook in another parking lot, and one of them on the parking garage. . . And our Air Force Chinook friends would drop off ice. They'd drop off a ton of ice; so we were like, okay, we've got to get it to where it's needed now. The City of Gulfport called and asked if we had any ice, and I said, 'Yes, sure.' 'Well, how much do you have?' 'Uh, about 2,000 pounds.' Complete silence on the other end. And I said, 'When we get our next day's delivery, we'll drop it off.' So I had Larry get a forklift to the pallet and drive it down to city hall. Nobody was there; so we dropped it into the middle of the street and left it. I did get a call from the mayor's office with thanks for the ice—'Sorry we weren't here, but we really did put it to good use.'

"I sometimes wonder what the chatter must have been amongst all the people flying over the community, the people who in some way touched us—dropped off a patient, picked up a patient. I wonder what our impact was beyond even what we can imagine. You know, if we had it, we were so distributive, so generous. Quite a few of them knew they were seeing something different here. You could see it on their faces; they were shocked. We provided a safe harbor."

§

Fewer than 20 miles west, the smaller Hancock Medical Center took the full force of Katrina's winds and storm surge that covered more than half of Hancock County. The Bay Bridge connecting the two counties was gone; roads and highways remained deep in mud and muck, hampering travel, and delaying the arrival of outside help. But two Disaster Medical Assistance Teams, DMATs, got to Hancock Medical Center Wednesday evening. Teams from Missouri and Florida arrived on-scene, allowing Administrator Hal Leftwich to close hospital services. Immediately, he turned emergency department operations over to DMAT and federalized part of the property. They quickly helped hospital employees sign up to work with the newcomers—all very well organized, Leftwich recalled, "except they forgot their air conditioners. The air conditioners were a couple days behind; so we scrambled for box fans to try to keep their tents cool. Obviously in South Mississippi in late August, it was pretty hot!"

Accustomed to seeing 800 to 900 individuals a month in their ER, he said, "We estimated that from Monday morning when the water went down to Wednesday evening when the DMATs actually started providing care, we had almost 800 people come to the hospital. And we didn't have any dry supplies; the electronics obviously didn't work; we had no power. But our nursing personnel were safe. They had their stethoscopes and assessment skills and took care of as many people as they could."

Regular disaster drills had prepared the hospital to survive and continue through the disaster. "It turns out that, when you need them, those drills—you do some things just out of instinctive reaction because you've drilled so many times. You know the right way to handle something."

Another big DMAT from North Carolina arrived later. Leftwich said, "While the first two were spectacular, and they set up on the hospital campus, we felt it would be counter-productive to concentrate everything in one area when the transportation resources were so limited. Having North Carolina down the street at what was a major highway intersection where people were coming to get their ice and their water and other supplies —it just seemed like a good decision; they agreed when we all discussed it."

Additionally, medical professionals from the Mississippi, Alabama, Kansas and Delaware Air National Guards erected a soft-sided hospital—known as an EMEDS, for Expeditionary Medical Support—and began taking 12-hour shifts to provide medical care and stabilize critical patients for evacuation to an established medical facility. EMEDS ordinarily provides expeditionary-force medical care in combat theaters—Iraq, for example. Mississippi Air National Guard Colonel Janet Sessums, senior officer on the ground, said this marked the first use of the resource, fully equipped and valued at $4.5 million.

"One of the interesting things from the standpoint of medical services was that after the first 10 days or so, one of the most in-demand services was dentistry," Leftwich noted. "If you're not able to brush your teeth and keep your mouth clean regularly, and you're drinking suspect water, there's a whole lot of abscesses that go on. The dentist was one of the busiest guys out there." The same held true for State Department of Health responders to Hancock County.

Beyond, further west from what became known as K-Mart General and established by a group from the Midwest, Hancock Countians could access what they called the Rainbow Clinic, which offered "a little different type of medical care and some holistic care, treatment of other than physical injuries," Leftwich said, "providing food, and doing a lot other services. They were another example of a group that came in and was very, very helpful. There was the free clinic from Virginia that set up in downtown Bay Saint Louis, another free clinic group set up in the Kiln, and another group set up down in Pearlington. Part of our challenge was just getting out regularly, talking to them, and making sure we all knew what each other were doing so that if we had a patient who needed a specific service, we knew where to send them and how to get them there. The ambulance service was absolutely wonderful, very good about getting patients where we needed to get them and making sure people had supplies." He credits Rick Fayard with AMR Ambulance. "I've got a video of him looking a little shocked. He said that at the EOC in Hancock County we were in command and in control of *nothing*. We just decided we were going to make the best of it, make sure people were safe."

Collaboratively, hospital staff and responders wrestled with federal regulations that might have hampered services. One issue: accepting donated supplies.

"We had trucks coming in from various drug companies with antibiotics and other things we needed," the hospital administrator said. "They were unable to accept it directly as a donation; so we developed a system with our not-for-profit foundation. The foundation would accept the donation officially and give it to the county-owned hospital, who would then give it to the federal government. It was much more stream-lined on-site than it sounds, but it pointed out a bottleneck there in logistics." In the aftermath, he would use that example to influence Homeland Security to develop a form private companies can simply use to assure donated materials have met all federal regulations and can be safely transferred for human use.

"The Joint Commission did a satellite broadcast," Leftwich recalled. "I did an interview with them, our nursing director, and our plant operations director. In that panel discussion with Homeland Security and different agencies, they said, 'Gosh. We need to fix that.' While politically it's popular to make everything look traumatic, the people we were exposed to were all trying to solve problems.

"I told one of the guys from the Army Corps of Engineers, 'I'm used to dealing with bureaucracy.' And I was humbled looking at the bureaucracy of the Corps of Engineers, but they still managed to find ways to help us get problems fixed. . . And you had agencies like the Coast Guard, whose helicopters were picking people up out of trees and off roofs and bringing them to the hospital and taking our patients who were still in the hospital—they were helping us transport our patients to trauma centers in Alabama. Of course, the Coast Guard also has the responsibility of dealing with hazardous materials. In a hospital laboratory, for example, and other areas you have some fairly hazardous materials that need to be very carefully disposed of, especially if the power has been off 16 days and they haven't been refrigerated—blood samples and all kinds of stuff. The Coast Guard did a spectacular job of coordinating the disposal. Some of those things, if they had not been properly handled, could have turned in to much bigger problems."

More than medical care professionals came to help in Hancock County. "We had a semi-truck pull up in front of the hospital," Leftwich recalled. "The driver got out and asked, 'Do y'all need some food?' I said sure. Then I looked on the side of the truck at the corporate logo for Firehouse Subs. He said, 'We realized how horrible it was; this is a refrigerated truck. We've got grills, we've got food . . .' and they set up right in front of the hospital and started feeding all first responders. They were here before most of the other assistance was here to help. This company was started by some firemen; they know what these guys are going through. They took their own resources, sent trucks over to the Coast, and started feeding people. How *cool* is that?"

The hospital administrator also applauds the local and corporate-level Walmart operations for particularly having helped his community. The sheriff coordinated with the Waveland Walmart, which took on more water than the hospital did, to provide security for people to retrieve for themselves and their families some of the canned goods and other items not destroyed. With a full assessment done, the store managers brought in a big tent, put pallets down, "and opened a small store right there in the parking lot and started selling things, many not normal Walmart items, that you needed just to survive in that climate at the time. They were helping people just get by. Also from the hospital standpoint, we had hospitals in other states send us relief goods; one in Florida wanted to donate some beds. Walmart picked them up and put them into their distribution system and delivered to our hospital at no charge. They have a large logistics system all over the country, and we were able to take advantage three or four times for large equipment that people wanted to give us. Their size and community nature at that point really helped us get going again."

Though some days seemed never to end, the hospital continued to hum and to advance toward rebuilding and recovery. That's where even Leftwich found FEMA frustrating. "Most of the time, they meant well, but the paperwork involved. . . You understand *why* they have to protect the assets we're asking for; it took us almost two years to get a set of approved plans and start putting the project out to bid, and three or four more years to build it. There are a lot of stories about how difficult that process was. The health care sector is hard to rebuild, and it's difficult to understand why you will rebuild and what you want to rebuild. The rules are that you have to replace like with like or justify why you don't. They seemed to understand that from a firetruck point of view, but they had a hard time making that leap with medical equipment. And some of it's you just can't find the stuff."

Dealing with the disaster, trying to avoid price-gouging, taking bids for work to be done, costing out re-use versus replacement of such as bathroom fixtures, and knowing that technologies change and capabilities to respond explaining to them what the issue is. So one of them says, 'Well, let me look at our tracking system. We've got a truck going across on I-20 right now that we can divert over there and bring liquid oxygen.' Another guy says, 'We've got a guy sleeping in Alabama, we can get him headed down there; he can probably be there in the next couple hours.' I'm just listening—wow! Could government do anything close to this? So, the next day not only did they service and get that tank back operational, they went in and stopped by—whether it was their tank or not—and made sure all those liquid oxygen tanks were refilled and oxygen was available at Gulf Coast hospitals without the first dime coming from state government, federal government, or anybody else. They did it because it was the right thing to do. . . checked, fixed, and serviced every hospital within hours. Needless to say, I wrote his name and number down, and every time I had a problem I could not fix, he was my best friend!"

§

Although the public health agency had planned and prepared, practiced, and updated their plans at least annually for years, Craig admitted surprises. "The real eye-opening part for me was that, while we've always had mass fatalities responsibilities, I don't know that as an agency we had thought *who* was going to do things like recover bodies. Identify the bodies, and then visit with the families about their deceased loved ones, and then transition them to the coroner. I guess we just thought the coroners' offices did all that. After Katrina, we learned one coroner was missing for the first week—they didn't even know if she was alive—and another was well-overwhelmed doing the best he could with the resources he had. And the expectations of Mississippians to take good care of Mississippians are very, very high. Probably of all my challenges, the biggest was doing that with as much dignity as we possibly could—we're not talking about people laying out, but *pieces* of Mississippians that had to be recovered. Fortunately, the Lord was with us. I didn't even want to contemplate the thought of sending public health people out to recover bodies; so I called Georgia Bureau of Investigations and asked, 'Will y'all come?' And they said, 'Yes, we will.' So we did an EMAC agreement through MEMA to bring in GBI; they did a great job assisting the locals with collections and identification through our own Mississippi Bureau of Investigation; they augmented our work force. And we called for federal DMORT to come. Once they got here—I can tell you I don't know how those people do the work that they do. But they are the most professional group of individuals I have ever had the pleasure of working with. And I think we took excellent care of the 238 deceased."

those transported in?' She said, 'Sure. No problem.' She said, 'We've got fuel bladders on board that are portable, and we can just put them on the beach.' We quickly got DEQ and EPA and said, 'OK, guys, here's your new mission; you're to go out there and figure out how to set them up on the beach.' She also brought in fueling tankers to use in proximity to Harrison County. They were not good for traveling on the interstates but could be used in some of the spots we had for military vehicles. At that time, we were getting more and more military vehicles. And we still didn't have a system for fueling. So that came out of the Iwo Jima Battle Group. I can't even remember what all I got from them, but they were giving us a lot of stuff. . . Nobody else would have told you this. Not many people knew that happened—Rupert and Bobby would remember; Joe Spraggins might remember. Rupert kept track of the logistics. Everything we brought in, he accounted for in the logistics process."

§

The physician at the World Trade Center on the morning of 9/11 and the first physician to the remains of the south tower that morning walked into the Harrison County EOC ready to work, ready to respond to whatever needs Katrina had created for the local medical establishment. Dr. Raymond Basri had been on vacation in San Francisco when Katrina blew into the Gulf Coast of Mississippi; he flew to Houston but learned they were overwhelmed with responders and needed no help. Back in New York City, the mayor offered his plane and medical supplies for New Orleans. Basri's flight—with two other physicians into Baton Rouge's private air terminal, with SUV waiting—would be his first of seven to benefit Mississippi.

The New Yorkers spent the first night in a hospital in LaPlace and the next morning drove east toward Gulfport, reportedly badly hit. They stopped at a community college just north of I-10, having been told volunteers were distributed from there. Go to Harrison County EOC, they were instructed.

"Then we saw the first major damage," Basri said. "I introduced myself to Bob Travnicek and told him my qualifications, and he immediately embraced me and put me to work. He gave us tasks to assess existing medical services. We pieced together information and learned about other problems, such as credentialing of out-of-state physicians so that prescriptions they wrote would be honored at local pharmacies. We had a problem knowing which pharmacies were open; we lacked insulin. And then we had a problem trying to monitor the new clinics or mobile trailers that had moved in to supply medical equipment and services.

"We took on each problem. Each day was a very long day. We participated in EOC meetings twice a day, at 7 am and 7 pm. We did what Dr. Travnicek thought he needed. We visited all the hospitals, and even went to Hancock County and tried to coordinate DMAT teams located in the parking lot of each hospital. We set up Thursday afternoon meetings for every week to discuss each hospital's actions and to learn about medical issues that might be a harbinger of an epidemic—dysentery or typhoid fever, for example. We made sure medical clinics going into the diverse local and minority communities were going to where they were needed; we did not want anybody to feel abandoned, but we wanted best use of all the resources. The first day we were there, we heard that 35 people had died of lack of insulin; that became a priority. I called Eli Lilly and had 1,000 doses of insulin flown in the first day."

A volunteer firefighter for 25 years in Middletown, New York, and deputy county fire coordinator for Orange County, Basri also is a senior aviation medical examiner for the FAA. He is a clinical assistant professor of medicine at New York Medical College in Valhalla, in private internal medicine practice in Middletown, and a diplomate of the American Board of Internal Medicine.

"I worked with Travnicek, and only with him, as the health commissioner for the Coast," Basri said. "Gulfport was entirely positive toward us from the moment we arrived. Dr. Travnicek is loved and respected. He has his own way of dealing with things—very conciliatory and inclusive. He has an excellent reputation with the political figures and the hospitals. To work with and for him was very easy.

"General Spraggins ran it like clockwork. The first week, a Jacksonville, Florida, emergency management team came in—they brought their entire staff, from top to bottom—as if it were their own disaster. They showed up a day or two before I got there and matched themselves with the local counterpart to become a shadow person, basically the same I did with Dr. Travnicek. We organized and supplemented the local teams so more work could be done."

Stories of a "shadow EOC set up elsewhere" reached Basri on his second trip to Gulfport. "They were making decisions contrary to the needs of everybody else, and rivalries were going on. I tried to ignore it because there was so much to do, but it did impede our progress to some extent."

On that second trip, he drove to Dedeaux Road to see for himself. "I walked through and saw it was a very substantial operation—and that's in conflict with every point of emergency management. Having a separate EOC for health issues is a fundamental no-no. It's a very basic contradiction of a disaster management plan. You cannot have that major entity segregated. It will set its own priorities and will not know what the group needs. It can't help the group, and the group cannot help it. They were also in control of many of the assets the group needed."

Basri also learned about the USNS Comfort's availability and non-acceptance.

"The ship's captain called me," Steve Delahousey said from his ESF-8 post at the courthouse. "Congressman Taylor and Senator Trent Lott had them come in, and we went over there. It's a Level One Trauma Center, but Dr. Amy and Jim Craig said that was not a state-approved resource."

On Representative Gene Taylor's staff, Beau Gex also noted the Comfort: "President Bush had a couple of hospital ships to come and locate off the Mississippi Coast that could transport patients who needed immediate surgery, certainly to an extent that hospitals couldn't. The one thing that was really, really bad about it was we had these two hospital ships ready and willing to help people, and this Dr. Amy up in Jackson who's head of the state public health group wouldn't let transportation of the patients go because he didn't think it was necessary. I know Gene talked to him. When he waved off the ships and said we didn't need them, Gene told him to not ever do that again!"

According to Basri's understanding, "The shadow EOC group claimed the ship would charge $1,000 per patient per day, and they threatened anybody who went there; so the ship pulled anchor and went somewhere else. Especially the Vietnamese fisherman and some other groups needed access to that ship, but we didn't even know about it while it was here. It was intentional negligence to not let that be used."

Jim Craig disagreed. "It's a great resource; it just got here too late. I think the intent was good and had it gotten there day two, day three, we probably could have used it. In fact, the first time I went down there, the Admiral asked, 'Is there anything else we can help you with?' I said, 'If you've got a fuel hose that'll run up to the beach, we could use some gas!' 'No—can't give up my fuel,' he said; so I told him, 'There's nothing else I can think of at this time.'"

MEMA Director Robert Latham said that by the time the Comfort could reach Mississippi, "the health care needs of the state were being met. We had free clinics set up; people had access to health care. The Comfort and what it was going to do was not needed, but it was deployed from Norfolk, came down the East Coast, and came into the Gulf. We kept pushing back on the Comfort, but there was this insistence that the ship should come to Pascagoula. We didn't need it, but we did say, 'Send it to New Orleans.' Emergency care and the free clinics and mobile disaster hospitals we had set up were meeting the need. The unified command evaluated every resource, and we came out of there with a plan. Not one soul in the UCG thought the Comfort needed to come to Mississippi."

§

To many Mississippians, Katrina stimulated a series of hits and misses. She missed New Orleans but hit Mississippi. She generated volunteers and donors from every state in the nation. Most of them hit the ground running when they arrived to help first responders, but some of them missed the point when they arrived unprepared. Steve Delahousey observed and commented: "We had volunteers from all over the country—actually there were too many; there was no coordination of the volunteer effort. And all the health care people coming in were supposed to come to the courthouse to get processed. Finally, the triage system worked.

"The faith-based groups were great. But others who came in here and started the conversation with, 'I need. . .' I said, if you say 'I need,' then get the hell out of my courthouse. Don't ask me for toilet paper, water, not anything; if you're coming in here saying 'I need,' then you've never participated in a disaster of this magnitude. Come in here and say, 'I have this to offer. . .' If you're coming in to help, come prepared to be self-sufficient!"

Delahousey developed extensive analyses of various response efforts related to ESF-8. He noted and spoke out about challenges, best practices, and areas for improvement. Among his biggest frustrations were the nurses and doctors who entered the EOC with "lists of controlled substances that they needed, and I'd say, 'What!?' Travnicek and I didn't have much help until we got the Board of Nursing here. As long as they were licensed in some state, everybody got automatic reciprocity—but many of them would say they had to go out to their car to get their license, and many of them would never come back. They were bogus—oh, yeah! They were coming down here to get controlled substances, to work the system. Keith Parker with the health department and I went out several times, and they were like crack whores; we would shut them down on the street corners. All these big pharmaceutical trucks were coming in, and they would see these little clinics set up on the street corners, and the trucks thought they were legitimate and would unload all these medicines that would never get to the hospitals. These people thought they were going to get the drugs. Keith and I would go and shut them down one day and find them on another street corner the next day.

"DMAT—I can't say enough about them," he changed gears. "They landed a day after the hurricane, and those guys got off the helicopter—Stan Crowe, the head for EMS said, 'We're a federal resource, and we're going to help, but I don't know this place. You tell me; where do you need these DMATs set up?' They came to help and asked us where we needed them. We set them up outside the hospitals—some of the hospitals just needed extra help; they didn't need the DMAT tent, just the help. They coordinated with the local people and hospitals and put DMATs outside the hospitals."

From inside the EOC, Delahousey and Travnicek conducted telephone triage with medical consultation. "Coordination of credentialing with the state EMS office—we finally made that process work really well. This was good. Normally, you had all the EMTALA laws to worry about: hospital administration, doing an initial stabilization, all that paper work, so much red tape. They told us that under a state of emergency, EMTALA and HIPAA did not apply. So any out-of-town transport, I immediately turned that to the state, and they did that well. I tell you who helped with that—I didn't care for him, but Mills McNeil did a good job with that. I told him my dispatchers didn't have time to do that. He did a good job doing what we told him to do."

§

Ray Basri and his team from New York validated Bob Travnicek in a way that nobody else could or did during the aftermath of Katrina. "I broke down," admitted the district health officer. "They gave me help. Basri invented the emergency response system in New York City on 9/11 and then came to help New Orleans, but they turned him down. So they came here, and he was the first person who was nice to me."

Travnicek credits Basri for suggesting a systematic approach to handling all the calls for help and for taking it another step toward pushing information about the availability of medical resources. "In New York, he had the idea for a central medical unit. The first day he came, he met with George Schloegel and Gary Marchand at Memorial. Basically, his proposal was to establish about 10 telephone lines with three or four nurses working 12 hours a day to provide hotline information. A couple blips after hearing the idea, Memorial Hospital of Gulfport agreed to the concept and established the service. They publicized the Memorial Medical Information Line through the EOC and advertised it in the newspaper and on television—'Call 228-867-5000.'"

A Memorial executive added, "We realized that people coming to the emergency room didn't know where anything was, particularly drugstores, because there was no central gathering of that information; so we set it up. People wanted to know where the doctors, hospital, and pharmacies were. Were they back? Dr. Travnicek asked if we would, and we turned out to be the information source for everything medically related."

Said another: "After the storm, people didn't know how to contact their physician or how to get in touch with any doctor, for that matter. They were out of medicine, and they didn't know how to get in touch with somebody to get medication so they would call. We got all sorts of questions. We had lists of all the doctors out of the newspaper and the telephone numbers. That was one of the problems—even when doctors were back working, the telephones weren't always working, or they had to get a new number so we had to provide that information. I think it was a good service. Those phones

Chapter 27

Public health workers covered Katrina on multiple levels, all important. From the Central Office in Jackson, at what MSDH called the regional ESF-8 command on Dedeaux Road in Gulfport, and at the county EOCs—every ounce of energy devoted to protecting the public health meant something to somebody.

The day of the storm, five years later, *forever*—Charles Stokes would affirm public health to be critically important in any disaster preparation and response. President and chief executive officer of the CDC Foundation in Atlanta, Stokes said, "There's no such thing as a disaster that doesn't demand public health services."

Michael Cruthird knew that. From his military service to his career as director for early intervention services in the Coastal Plains Public Health District, he knew the urgency for response to community-wide public health needs. Even though he suffered at home through a bad bout of gout before the storm, he helped organize emergency operations in Stone County. After Katrina's landing and with most of his personal concerns satisfied—his and family's safety and belongings were intact, for the most part, and dear friends Jennifer and Doug Handshoe with son Eric had survived roof-surfing in Gulfport—he went back to work Thursday. Jim Craig had sent a Florida Department of Health responder to assess facilities in Stone County, where Cruthird lived. "I've been shot at, and it never affected me as much as this storm did," he admitted. "We traveled the roads all over Stone County, checking on the status of the facilities, on their emergency power supplies, the diesel, the food, on the dialysis centers and trying to figure out what we were going to do with all the dialysis patients. We ended up setting up a dialysis center over at Pearl River Community College.

"When I came back to the incident command center and made my report to Jim, one of the things they told me at the EOC in Wiggins was that we've got people calling here for baby food and diapers, just wanting milk or formula. And that's when I realized we had a warehouse full; so I called Jim, and I said, 'Jim, I'm not asking permission, but I will be asking forgiveness.' And he said, 'The governor said to do whatever it takes. If you can't get the keys, break the door down.' Well, we got in with the police, and we took items that the county needed and put it into the hands of the County Extension Service home coordinator, who was helping with relief supplies to the whole county."

When he was able to get to the district office in Harrison County, Cruthird agreed to Administrator Kathy Beam's request that he be local liaison for the regional ESF-8 command, also known as EOC-North. A partial and wide-ranging "sample" day's task list included checking on tetanus vaccine for Biloxi and Singing River, checking evacuation plans for the shelters, fly and mosquito control issues, ask Illinois for standardized forms, FEMA packs, get WIC back open, get gas, put together a pharmacy and distribution plan. He also took responsibility for helping re-establish public health services in Jackson and Hancock Counties, whose facilities Katrina had totally destroyed.

"I worked a lot of hours," he acknowledged, "and we had some real public health heroes who also did. I called some of my people and said we're not doing Early Intervention but if you want to work, there's work to be done. My administrative assistant Rita Brock and one of my service coordinators, Jennifer Handshoe, said, 'We have nothing else to do right now'—so they came to work with me: taking phone messages, making copies, whatever needed to be done administratively at the command center. Jennifer worked every day that she could work; we were the local presence at the center. Everybody else was from other parts of the state. . . You had nurses and clerks who came to work every day in the heat and humidity to work under the shade of a tent and—if they were lucky—to have a box fan and MREs or Salvation Army lunches. And thank God for the Salvation Army—Red Cross I won't even talk about—Salvation Army gets my money."

"Our public health people who deployed to the Coast were in a stressful environment. The district office was in bad shape, even though they later established emergency power, and they used it for housing for some of our personnel from north Mississippi. We could not buy fans because they would not let us go to Sam's. We needed tents. Sam's had them and said, 'Come get them; you can pay for it later.' Jackson, the Central Office, said, 'You're not authorized to shop at Sam's.' Various people rotated in and out, and when the purchasing director finally brought his butt down here and spent 24 hours eating cold turkey sandwiches, sleeping on a cot in a warehouse, and going without a shower—all of a sudden, purchasing got a lot easier!"

"The WIC Center became the big operations command post, Regional; they worked there, ate there, and finally brought in portable showers, and that's where the 'club house' was. They brought in that big travel trailer that you got invited into if you were a member of The Club. It was a nice mobile home, and it had air conditioning."

Fellow public health workers, according to Cruthird, fit into one of two categories based on the work they did and how they spent downtime: unsung heroes and The Club. "There were good people at Regional," he said. "We had sanitation people down here who knew what they were doing with wastewater. They had as little to do with Regional as was possible. They used The Club for information, resources, and to get a hot meal, and that was about it. We had good nurses down here, but there were limits as to what they were allowed to do, too. They were not in The Club, not members of 'the club house.'

The Club, he said, were "the black shirts—Keith Parker, Art Sharpe, anybody associated with EMS and disaster response" or who was "invited" by those employees to participate in questionable after-hours activities —"incidents that shouldn't have happened. There were plenty of mature people there who knew what they were doing—can I say that without disparaging them? I don't want to do that at all. Theirs was an uncomfortable environment—hot, sleeping on cots, not getting to go home and have a hot shower and clean clothing. They had to relieve some tension, but I think some could be accommodating and were abused. Art always had a lot of swagger. Art was more into taking tours, wearing his sidearm, going out to

Keesler to get hot showers and good food, and shopping at the PX. A lot of us would love to have taken tours, but we didn't have government vehicles and couldn't get gas. Even though there was a tank of gas sitting over there, it was reserved for official vehicles, and if you weren't a member of The Club or given a specific task, you were not allowed to use those vehicles. All I know is that I never got to be a member of The Club."

MSDH's After Action Review documented workers' contributions based on categories outlined by disaster health and medical priorities: planning, personnel, pharmacy, physical plant, and product. The agency mobilized more than 1,400 non-disaster personnel to fill vital response roles for some 10 weeks. Most individuals got just-in-time training to handle their new roles. "We were real fortunate we did not have an infectious disease outbreak," Cruthird said. "One reason, one big factor is that Mississippi in Katrina successfully deployed and received the Strategic National Stockpile. That was huge. It had never been done before and has not been done since.

The department successfully requested, received, staged, and distributed the Strategic National Stockpile, making Mississippi the 'first state in US history to complete the entire cycle prescribed for the SNS and providing pharmaceuticals for the state.'"

As they historically had done, public health workers focused on such issues as water safety, private water wells, waste water, and food safety—re-opening restaurants. They supplied nurses to clinics, provided immunizations, mitigated vector issues, assisted with myriad assessments, and attended to special medical needs populations. Cruthird could say that what was happening to Travnicek was not really that much different from previous disaster response events, "but he felt under fire. It was a huge effort on his part. Harrison was the focal point, the population center for so much."

Cruthird found himself straddling a wide gap between working for the EOC-North command staff and fulfilling what he knew to be vitally important relationship issues that could meet his community's primary care needs. "What was going on in Jackson, I have no idea," he said. "I only know that Dr. Amy came once. He showed up at the ESF-8 command and insisted that Travnicek be there. I hated the way he was treating Travnicek because I knew what Travnicek had been going through. They told me to call Travnicek and tell him to go home because he was 'not talking good sense' and was 'being an irritant to the good people in Harrison County.' This was coming from Art Sharpe. I called Travnicek. There is no doubt in my mind the man was suffering from sleep deprivation. I asked him if he would go get some rest; he accused me of being a tool for Brian Amy. He knows better. I reminded him of that, and he apologized and said, 'Michael, you have to understand: if I leave here, I'll never get back in. They'll send somebody else in here that will claim I've been relieved. I cannot leave.' He promised me he would get some sleep, and he sounded a lot better the next day. That's when I figured out, he probably was right. They did want him out.

"Amy ordered Travnicek to come to the ESF-8 command center and then made him wait for hours before Amy showed up. When Amy did arrive, I made a point of going on up to them—he and Travnicek were talking— and asking Travnicek to review and approve something. I forget what it was, but it was important enough to make an impression on Amy. I did it just to emphasize to Amy that Travnicek had friends and had people who were working for him down here. Amy got the message. I don't know that Travnicek knew what was going on, but when Amy got ready to leave, he came by and told my people what a great job they were doing and how much he appreciated what they were doing, and he snubbed me. He didn't just bypass; he went around and away from me—did not look at me, would not shake hands with me. Not that I was going to shake hands with him. He was not one of the heroes.

"Travnicek was. He made hard decisions, and he did some things that only a true public health officer could do. Can the man rub you the wrong way? Absolutely. Can he get on your nerves? Absolutely! But Travnicek was well connected, and Amy was *scared* of him."

Cruthird put Travnicek and Delahousey in the "heroes" category with Joe Dawsey and Jeff Bennett. "The heads of public health, emergency response, federal clinics, and mental health down here did a great job; they were heroes. They pulled and put things back together the best they could." Collaboratively and under the ESF-8 unified command, Cruthird advanced the Interim Primary Health Care Plan, aimed at rebuilding a healthy community through integrated medical and mental health service delivery. In the document he submitted September 15, 2005, they envisioned restored public and mental health services for the six southern counties to at least the same level provided before Katrina. Additionally, they suggested, the plan "should help revitalize the economy of the Mississippi Gulf Coast region by restoring and sustaining employment in the local area's healthcare industry."

The three-phase plan could be completed in 180 days and would use on-site federal assets, particularly DMATs; discover customers to be comprised of residents, recovery workers, contractors, and visitors; inventory available primary health care providers, including private and public practitioners; determine physical resources; and inventory all personnel, equipment, and physical structures remaining and expected to be available within the six-month period. Other plan aspects covered public information, an interim primary health care plan project manager, HMO/hospital plans, survey teams, benefit providers, pharmacies, and transportation. For Hancock and Jackson Counties, whose public health facilities were gone, the proposal called for the lease-purchase of two GE Modular Space (GEMS) units, each designed to meet space needs for the area's anticipated population. Projected cost for the installed units totaled less than $1 million.

Regional command staff approved the plan that day, September 15, and sent the proposal to Jackson for the state health officer's signature.

"I go back and look at that plan every now and then and think, God! That was an opportunity we lost," Cruthird regretted. "We pushed that idea, and pushback came through Amy's office. I don't know if he ever saw it—I think Danny Miller did. But this provided for one-stop shopping for healthcare. Essentially, you presented yourself, and you would be triaged; depending on your needs and circumstances, you would be seen by public health, mental health, the federal clinic, the private physician or all of the above or any combination thereof. And there would be an integrated patient-management system to provide accountability at the unit level so that public health was still public health, mental health was still mental health—you had your numbers, expenses, revenues, and reimbursements. It was approved, but never implemented. I think Art bought off, and it went to Jim. I briefed him, and I don't think he disagreed, but Danny Miller ultimately said, 'We're not concerned about mental health and the federal clinic; they can take care of themselves.' That's when I asked Jim, 'What the hell is ESF-8 supposed to be doing?'"

Despite such distractions and disappointments, Cruthird continued his quest to reconstitute healthcare services. "We had a lot of volunteer doctors flooding into the area. They were not signing in but just showing up and establishing street corner clinics. By late September, Memorial was concerned that the free clinics were providing uncompensated care and taking patients from local doctors. Memorial was talking about having to build a housing area just for their own staff; they were concerned about losing patients and losing doctors—physicians were threatening to move." As liaison for public health between EOC-North and Harrison County EOC, Cruthird crafted a news release distributed September 22 and styled as "Memorial Hospital, Department of Health, and Unified Command Request" for a meeting with all temporary emergency medical units. All such temporary EMUs in the six-county area were invited to the courthouse the next day at one o'clock with the promise that all could "discuss unfulfilled medical needs within the region and the provision of the primary medical services."

Although the meeting did occur, Cruthird "got read the riot act from Jackson because I was 'circumventing the state's procedures.' They perceived that we intended to threaten volunteer doctors—there's no threat implied in that! We never made a threat. We simply said, 'You must register with us.' We were interested in being sure that we protected our local medical assets." Two years later, he would note, the department "asked me to go check one of those out. What? You read me the riot act about this, and now you want me to go spy for you?"

Jim Craig now talks about the importance of the recovery process. Before Katrina, he acknowledged, the department's plan did not adequately recognize chronic illness support needs—"We are trained to stop the bleeding and start the breathing"—and did not rapidly address helping health facilities and practitioners get back into practice. "We lost many physicians. We should have been more clear about the engagement of health care." The agency's recovery framework now addresses those important issues.

Chapter 28

Hooty Adam's not big on flying, but he did a helicopter surveillance of Katrina's damage to Hancock County on the second day. Unbelievable—that's the one word he uttered as he tried to determine their location. "I'm from Bay Saint Louis, born and raised there, but when I got up there, I could not recognize anything or figure out where we were. I just kept saying unbelievable. And when I got off the helicopter and started talking to our group at the EOC—they said, 'You kept saying, unbelievable, unbelievable.' It was. Absolutely. Unbelievable."

With County Administrator Tim Keller, local law enforcement, and military officers, the EOC team planned Monday evening for their work at first light Tuesday. "There was so much to do and not enough people to help," Keller said. "No communication. No phone lines. People were coming in and telling us they had seen bodies. We realized we had to find out where they were. Others were reporting to us they could hear voices under downed structures. I still remember those. It was just a crisis mode."

Adam remembered how it started. "Brice Phillips was here. Brice operates WQRZ Radio and founded the Hancock County Amateur Radio Association. He moved into the EOC before the storm and stayed with us the whole time. He was a tremendous asset for not only the EOC but for the citizens of Hancock County."

Search and rescue started. Law enforcement was to take care of bodies.

"The third day after the storm," Adam recalled, "we're in the back of the EOC and a group from Florida, an EMA director and Incident Management Team from Manatee County, showed up. I'm also fire marshal for the county, so I carry a pistol, sometimes; I had it on my side. When she introduced herself, Laurie Fagan said, 'Oh, EMA directors carry a gun now?' And I said, 'No, that's just in case y'all get out of line.' We didn't know they were coming. We had no clue. But let me tell you this: they were life-savers. Their help and dedication to us was just unbelievable. Unbelievable."

The Florida group gave Hancock Countians the first evidence that people beyond the county line knew they had survived. Even though advised to evacuate before Katrina's landfall, Adam and other stalwarts stayed. "We kept getting calls until the phones went. People were stuck in an attic, floating down the road in their vehicle, and then silence. No communication. No nothing. For days," the EMA director said. "We were thought to be dead; my wife and three boys had evacuated to Tallahassee and heard on television that Hancock County's EOC was gone. She almost had a stroke. My boys were upset, and they immediately took off to get back, but it was two or three days before I could even see them. Roads were impassable that first week. You had to cut your way through to most of the places. After the water receded, we found a boat that ended up on a house on stilts, 15 or 20 feet up; that's how high the water got."

EMA director in Hancock County since 2002, Adam earned quite the reputation for the way he handled EOC meetings. One observer from Mobile commended him: "They say you talk the way you see it."

He acknowledged the perception with self-confident pride, and a sense of humor. "We do things a lot different than Harrison County and Jackson County. Rupert and John Albert and I—all of us are great friends. But Rupert does what Rupert does, and Hooty does what Hooty does. I'm a lot more unorthodox than most people. I allow some media in; Rupert will not. I tell him all the time—I'll cut a joke. When we started getting these IMT teams in for Katrina, we were at the Vo Tech center by the airport, and I found this little school teacher's bell—ding, ding, ding! And everybody would introduce themselves, and I'd hit that bell—ding! Then real quick, I'd say, 'That's enough. Sit down!' and they were stunned. Then I said, 'Nah, just joking.' But you have to break the monotony. It started with Eric Gentry and Colonel Melton, with the National Guard. I'm telling you— there was no smile, nothing for days after the storm. So we started playing jokes on each other in our meetings, and that started relaxing everybody.

That's how we dealt with it."

§

Early Tuesday morning, FEMA representative Eric Gentry assured the team that disaster supplies for Hancock County had been ordered and were in the pipeline. Congressman Gene Taylor told him, Keller recalls, "I want you to see that they get here, and I want to know when supplies get to Stennis. I want it tracked."

Taylor shared Hancock Countians' expectation that water, ice, food, and other supplies would be delivered Wednesday morning. "Where are the supplies?" he asked at the Wednesday meeting. Gentry could not confirm. Keller said Taylor "actually put his hands on Eric Gentry and said, 'I want to know. I will have your ass fired if those supplies don't get where they're supposed to be.' After he shook him and got his attention, Eric came and said, 'I've got to have this congressman off my back so I can figure out what's going on.' So I personally went and said to Gene, 'This man's trying to help us. I respect you up on that federal level, but trust me on this; at this level, this man's trying to help us all he can.'

"Eric made that his priority, went and locked himself in with satellite phones, and then he came back and reported. He said, 'I found out what happened to the supplies. . .' Now these were *trailer loads* of supplies; I'm not talking about boxes but several trailer loads: MREs, ice, water, all the essentials. Eric came in and reported to us—he got to the bottom of it —and said the Harrison sheriff, the chief law enforcement officer, had gotten wind that they were coming to be staged in Hancock County and sent his deputies over to Hancock County and ordered the drivers to come to Harrison County with all the supplies. They actually commandeered federally-directed supplies that were coming to us and started disbursements. Gene found out later that day, and the supplies made it to us the next day. We were a day behind because of that. And the sheriff in Forrest County did the same thing."

§

Among first actions, the Florida IMT set up a press conference for Wednesday afternoon. Media had not been on the locals' priority list. Command staff had started spotting bodies Monday, immediately after

Katrina's eye came up the mouth of the Pearl River and battered the entire county with her wrath. "We were in the worst part of it, on the immediate east side," Keller said. "Monday night, a man by the name of Chuck came from the hospital. I'd never seen him before and haven't seen him since. Chuck had been sent by Hal Leftwich to report that there was a casualty in the hospital. They didn't have any way of preserving the body, and they came over to see what we could offer. We went over to Mississippi Power and loaded up an ice chest full of ice and sent it over so they could ice down the body there on the ward on the second floor." They later discovered the deceased to have been a former justice court judge.

Keller and Adam started the scheduled news conference with a tightly scripted presentation the Florida team gave to them. "And then Hooty just started talking. And after, Laurie Fagan said, 'Don't let anybody here upset you. You've just opened the world up, and you've touched everybody in this country with the stories you've told.' I think she was pleased we went off script."

Keller thought about what he would say the next time he addressed the media—at the invitation of General Spraggins at the Harrison County EOC. "That was our first time to touch national coverage. When it was my turn, I said, 'I'm reporting to you from Hancock County, Ground Zero in Hurricane Katrina' – because by that time, everybody else was trying to get all the attention."

For National Hurricane Center Director Max Mayfield, Hancock County claimed the attention, and he went to visit. "I rode around with him and Eric," Adam remembered. "And he is one of the most genuine people I believe I've ever met. Max Mayfield is wonderful. He absolutely took to heart our suffering."

§

Several three-ring binders contain page after page of eight-by-ten photographs Adam and other responders shot as they conducted search and rescue. One shows an open Bible, wet, with pages beginning to curl. Beside a cypress tree and near a blue blanket on a bed of pine straw, the Bible delivered a promise.

"We probably had the worst debris pile," Adam displayed. "Let me show you. This debris pile, that's Bayou LeCroix. It's a large, large area, and this area was filled about four, five, six feet deep with debris: houses, with everything. I don't know how many acres, but it was very large, and I found this Bible, opened up to the verse about Noah's flood. Eerie, very, very eerie."

¹² God added: "This is the sign that I am giving for all ages to come, of the covenant between me and you and every living creature with you; ¹³ I set my bow in the clouds to serve as a sign of the covenant between me and the earth. ¹⁴ When I bring clouds over the earth, and the bow appears in the clouds, ¹⁵ I will recall the covenant I have made between me and you and all living beings, so that the waters shall never again become a flood to destroy all mortal beings. ¹⁶ As the bow appears in the clouds, I will see it and recall the everlasting covenant that I have established between God and all living beings—all mortal creatures that are on earth." ¹⁷ God told Noah: "This is the sign of the covenant I have established between me and all mortal creatures that are on earth."

–Genesis 9:12–17

§

Hancock County mirrored Harrison in the aftermath but, surrounded by water, was isolated in the midst of total devastation. They relied on what they had.

"I don't remember where he found it," Keller said, "but Boss Hog had found a cook-top with four burners on it. We started realizing that everybody had come to the flooded EOC, and we got into the food that was on the higher shelves of the food pantry and into the freezer. We set up a cook tent, a tarp, and you won't believe this, but Boss fed between a thousand and 1,200 meals a day—not three meals a day, but for the next five or six days, from the four-burner stove, we'd cook grits, eggs, wieners and bread. We had a bunch of bread. Nothing real nutritious, but it was food."

Food stored at the EOC was intended for emergency management personnel and law enforcement. "But people were walking in off the streets because they hadn't had anything to eat," Keller said. "I remember telling Boss, 'We're going to feed them till the food's gone because we can't turn away hungry, thirsty people.' They didn't have anything. A lot of them walked up with small children, came on bicycles; all of them had horror stories."

The EOC command staff regrouped, relocated, and welcomed help from Emergency Management Assistance Teams from Florida; they welcomed the military and volunteers; they entertained and tried to provide for the homeless; they shared sandwiches with dignitaries such as Former President George H. W. Bush.

"We were Ground Zero as certified by National Hurricane Center," Adam declared, "but we're also the land mass between Harrison County and New Orleans." Knowing that he and his team had survived from the storm's epicenter but had remained largely invisible to the rest of the world annoyed Adam. He struggled to secure attention of outsiders who would contribute to Hancock County's recovery, and he knew President Bush could. He also believed that some of the people who came had their own agenda—"we had a lot of cowboys coming in."

Neither Adam nor Keller liked that a high-ranking Mississippi law enforcement official breached what was supposed to have been a private meeting for Hancock Countians with Bush, "just for Hancock County to discuss Hancock County issues," Keller said. Suddenly an imposing, uniformed man "just came in, flashed his credentials, and tried to emphasize law enforcement," Adam interjected. Keller continued, "At that time, we were days into this thing, and not much diplomacy was taking place. Everybody was nervous about that. But finally, Eric Gentry got it back under control, and we were able to tell the president our story."

§

Then and now, Adam affirms that Hancock County "got lost in the shuffle; I don't care what anybody says.

"But we got one of the best things after the storm because New Orleans didn't want it, and that was North Carolina Med, a portable hospital. New Orleans wouldn't let them in. Our hospital was knocked out, and we were trying to find something. They called us and wanted to know. The military was trying to take care of things with a field hospital, but we needed something a little more permanent, and we just happened to have a guy by the name of Jeff Cartwell who worked for the North Carolina Emergency Management at the State level. And I turned to him and said, 'What do you know about this North Carolina Med?' He said, 'Bubba, if you can get it, grab it.' He said, 'They're A-Number-One'—and he was right."

§

City of Bay Saint Louis Attorney Donald Rafferty arrived Thursday. Born into a family whose Pass Christian home Camille claimed in 1969, Rafferty and his immediate family survived Katrina with his brother and his family, also in the Pass. Katrina spared both homes with relatively little damage, despite the duration.

"Camille was in and out in four hours, and we thought she was the devil of all devils; however, we were surely wrong on that," Rafferty compared. "The biggest difference between the two hurricanes is that Katrina just languished here and beat us up for almost 24 hours." From late Sunday afternoon until the middle of the next day, well into the afternoon, Katrina was like "a 24-hour beat-up compared to six hours for typical hurricanes when they came through. We were so unbelievably lucky. And the next morning, people were out cleaning up, working, helping each other—instead of going on roofs and bitching at the cavalry—until the cavalry arrived, and then we were grateful."

Living and working on both sides of the Bay of Saint Louis gave Rafferty insight into both communities and catapulted him to become a courier. He had to go the back way, get from near the beach in Pass Christian onto I-10, drive among debris that had not been cleared, and go south on Highway 603, with only one lane open because portions remained underwater. What normally would have been a half-hour drive became nearly an hour-and-a-half.

Arriving Thursday in the City of Bay Saint Louis, he went to what's now the center for the city, the VCJ Building, the fire department, which had become the city's internal command center. From there, Rafferty went with the mayor, buildings officials, public works staff, and fire and police to meet with Hancock County emergency management officials. Daily meetings for the first couple weeks focused on search and rescue, search and recovery, and then what the city needed to do to get back on its feet.

"We had our first city council meeting the next day, on Friday," Rafferty recalled. "And that afternoon, the buildings official and I myself went down and actually engaged the CSX railroad bridge company—the company that was rebuilding the bridge—and we had to caution them on proceeding any further with the train bridge without coming to get a permit, even though we were not in a position to issue one. But within an hour they would have a very large check to pay for the permit before building the train bridge back up. We were already conducting regulatory acts this Friday afternoon after the storm to assure that we were all in compliance with the rules and regulations, to make sure it was done properly. They paid approximately $19- to $20,000, which on the Friday afternoon after Katrina was a large sum of money just for the permit on the Bay Saint Louis side.

"At that time," he cautioned, "some people didn't want the railroad to rebuild; they wanted them to move north of I-10. So here they are out there, the private company who has the money to do the work, not worried about anybody else, just trying to rebuild so that if, in fact, somebody tried to move them. . . You can connect the dots. I'm just telling you that on Thursday, we noticed there was construction work—a company called Scott Bridge Company—and nobody was trying to remove debris and clear it away. They were there to rebuild the bridge. That's significant because we're trying to save lives, find lives, and recover people. I'm just telling you what the difference between government and private enterprise is. You connect those dots; you interpret them how you want. Their concern was about business and getting back up and going."

Even though the Bay Bridge connecting Hancock and Harrison Counties would not fully open until January 2008—two lanes opened in May 2007 at an expedited rebuilding cost of $267 million—the railroad bridge's reconstruction would require only months. Whether those facts can be considered positive, Rafferty said, is "up to interpretation. On the one hand, yes. On the other hand, as people say, they came here with $24, beads, and trinkets and got our fine property, and now they go right through the middle of it all day, all times of the night, and stop traffic and everything—all for their commercial development. You can interpret it however you like."

An obvious observation centers on communications. "Cell telephone service was very restricted," he recalled. "There was one spot where the phones had reception: at the foot of the Bay Bridge on the Bay Saint Louis side. You'd see people congregate there every evening so they could get the word out with family and friends as to what's going on. It was like a rallying cry. And over at Pass Christian, the rallying point was at a FEMA tent on the former baseball field."

Rafferty's daily routine started with meetings and news briefings at the EOC in Gulfport, where he would deliver public information for the City of Bay Saint Louis about "curfews, FEMA trailers, food pantries, clothes, volunteers, and things of that nature. I was very fortunate to be able to sit in on both Hancock and Harrison County meetings of the mayors and powers that be, privileged to have contact with both counties and serve as sort of a liaison for both, at least to get the word back and forth."

Most conducting-business days included drafting invoices, receipts, and other documents on legal pads and sometimes passing hand-written notes between officials in the two counties. Everybody knew that government money for rebuilding their communities would come only with documentation. Katrina had crushed their infrastructure, not only destroying surface buildings and bridges but also obliterating the underground water and sewer lines; nobody could survive long or well without potable water and adequate sewage disposal. Those responsible for doing the hands-on, back-breaking work to handle those needs draw highest praise from the city's lawyer.

"Ronnie Vanney, public works director at the time, and his number one lieutenant Buddy Zimmerman need to be complimented," Rafferty said. "They went in and did all the dirty work, clearing the roads, taking care of the debris; they busted their tails—they and their crew of about 25 people. The unfortunate part is they didn't get the recognition that the fire and police departments did—I'm not trying to diminish the fire and police because, to their credit, the firemen and police were the ones out there trying to save lives. Once the storm calmed down, the police were trying to keep people from looting, keep traffic going; the fire department was out trying to rescue and save people; and the public works people were out there pushing away the debris with equipment that had been almost totally destroyed, knocked out, wet. It was just unbelievable what they were doing with equipment they had. Along with the fire and police departments, they were out there clearing debris, moving buildings, and trying to find people and save lives."

Health and medical care teams also worked to recover bodies, treat the injured, and make sure people got their tetanus shots. As in Harrison County, officials also dealt with access into demolished areas. "Most important," Rafferty said, "we knew a number of deceased people were out there, and at what point do you let people back in? How do you keep people out? At least on the west end in Bay Saint Louis and Pass Christian, it was a balancing act, but it balanced out well, having the ability to protect the public, to make sure no issue could expand due to disease or other things. I think we did a good job from a martial law standpoint, making sure that people who came in had a legitimate reason to be there."

§

Determined through the worst of Katrina to man their stations at the EOC, Adam and Keller knew they could not stay there for recovery operations. Keller said that by Friday, the facility had "become a pigsty, filled with wet carpet and wet sheetrock. Dirt, people with soiled clothes, and the stench of sewage made it almost impossible to stay. We looked at the Bay Saint Louis ballpark, but there were no buildings, no water, or sewer. We looked at the Hancock Vo Tech Center off Highway 603, then went to the School District Office, met with the facilities director, and asked to check out the center; it would be a perfect fit." They moved Saturday.

By then, Keller said military personnel began to arrive with "lots of volunteer civilians—Kansas, Florida, South Carolina, California—the word had apparently gotten out to the general public that help was needed at the EOC. Herds of needy citizens continued to come to the EOC. Everybody was coming—the word had gotten out that they could get fed there, get water. The emergency support function was developed to help us send them to where they needed to go for assistance. Almost overnight it became a town. Just like that—it turned into another town. People on the water towers, setting up antennas and satellites, military people. . . Within hours, there were tents popping up in every space on that whole campus: military tents, search and rescue, recreational vehicles, motor homes, campers were being set. Communications satellites were being installed, lines were being run to all necessary rooms, and a team of people came and set up a water tower and transmitters. It was amazing to see all the transformation for it to become fully operational."

Beyond President Bush, responders included Homeland Security Chief Michael Chertoff, Admiral Thad Allen, Under Secretary for the Department of Agriculture Tom Dorr, Lieutenant Governor Amy Tuck, then-State Auditor Phil Bryant, senators—"it just went on and on and on," Keller said. "Within one week, I estimated there were three to four thousand people on the school grounds, most of them military. We had the Buckeyes, that satellite group. We had more National Guard than you can shake a stick at."

But the local sheriff, said Adam, "was nowhere to be found. Finally, we got him there on Thursday. He is the chief law enforcement officer, but he still had to operate under the chief emergency management officer, the incident commander. He did not have any control; it took a while for him to work with us."

Keller commended Adam's self-confidence and steadfastness as incident commander—Adam, after all, was the local leader who did not see his own home for more than four months after Katrina's landfall: "After about 10 days of this thing, I was able to get away for a little while; he didn't. My story goes for 10 days. Hooty did the heavy lifting and stayed firm: 'This is the way it's going to be.' It could have been one of those cases that anybody who shows up and declared himself to be is the captain of the ship, and a lot of them wanted to do that. But Hooty kept everybody together."

Adam had refused to evacuate before the storm, and, never the martyr, he stayed after. Instead of accepting Harrison County's invitation to participate in news conferences at the Gulfport EOC, Adam sent Keller. He respectfully declined the request to testify before Congress in Washington, DC, instead maintaining his post as emergency manager. "I had a job to do."

§

Hancock County First Responder Brice Phillips also refused victimhood.

"I am disabled," he declared. "I have ADHD and long-term depression. But when I manage my meds . . . I can pull something out in very short time. This is why I'm a first responder. That's what it's about. I'm just another part of the team, and that's an asset."

Phillips organized Hancock County Amateur Radio Association, Inc., which operates WQRZ-LP 103.5 FM and WQRG 96.3 FM—"Serving our community through the 'art' of radio since 1994." The community encompasses Bay Saint Louis, Waveland, Diamondhead, and Kiln, and WQRZ-LP is 100 percent constructed, managed, staffed, and run by volunteer persons with disabilities.

Phillips' heroism kept WQRZ on the air throughout Hurricane Katrina's monstrous assault on his community, and after her catastrophic devastation, WQRZ staff communicated life-saving information about safety and health, points of distribution for ice, water, and other commodities, and answered questions from listeners. The station took many AMR calls, helped Hancock Countians contact family across the nation, and helped get parts shipped in to bring the Bay Saint Louis water system back up.

Of more than three dozen radio stations along the Gulf Coast, WQRZ was one of only four—and the only one in Hancock County—to never lose power and to stay on the air around the clock. Almost immediately after the storm, FEMA distributed some 3,500 portable radios county-wide so citizens would have access to vital information. Small Business Administration honored Phillips in 2006 for Outstanding Contributions to Disaster Recovery by a Volunteer.

"When all else fails," Phillips said, "amateur radio works until other systems get put in place and put online. I didn't have time to think about it. It was time to hook another radio up. When anybody is left with the last resource, you share it with your friends. You share it with your family. You share it with your community."

Phillips shared with his community and, in time, was officially named public information officer for Hancock County EOC and also was awarded a FEMA public assistance grant under terms of the Stafford Act to his non-profit radio operations.

"This is what I created," he said with a big toothy smile of pride. "I'm a thinker, and I work directly with the public. I wrote my own incorporation papers. I did my own 501c3. This is a first. No other radio station has an affiliation with their local EOC as WQRZ does. I'm eccentric. And I'm the model."

"Many of these agencies were completely demolished. Buildings and equipment destroyed, communications systems disrupted, and employees facing personal challenges that made it difficult for them to report to work. Despite these challenges, public health teams throughout the region were still desperately trying to meet the health needs of the communities they served.

"I started asking people," Stokes later told Travnicek. Invited along with the heads of Red Cross, Robert Wood Johnson Foundation, and others to tour the Mississippi destruction and do a needs assessment, near Bay Saint Louis he visited "a large building where the Red Cross had set up a major station serving food, providing shelter and health services. I distinctly remember walking up to the Health Desk—a folding table—behind which stood a volunteer nurse who had driven from Florida to help the Red Cross. Because I still had doubts about whether we could help with buildings, I decided to ask her opinion about the relative importance of replacing the local health department building in the midst of so many other needs. She dropped what she was doing, ran around the table, gave me a hug, thanked me, and said a new health department building was exactly the highest priority. She explained that the heath department in Bay Saint Louis was working out of a tent in a K-Mart parking lot and was open only part of the time. She said that Red Cross nurse volunteers, for liability reasons, could not provide certain services such as immunizations and, therefore, routinely sent people to the health department, and those people often expended precious fuel driving around downed trees and other obstacles to a health department which was closed. She said you desperately needed a new building from which you could begin to serve on a regular schedule. She convinced me that your request was real and necessary, and we then made the decision to fund your request."

Harrison County's health department in Gulfport, built to withstand Category 5 hurricane-force winds, had survived but flooded; it could be salvaged and returned to service. But when the eye of Katrina passed over Waveland and Bay Saint Louis in Hancock County, the health department building was completely submerged. In Jackson County, the roof of the Pascagoula health department building was blown away, and the building was drenched in rain water. When Stokes and Travnicek first talked, both structures were filled with debris and sat molding from the inside out, along with most of the furniture, equipment, computers, and patient records.

Immediately after the storm, the Hancock County team temporarily worked in a mobile clinic—essentially a large van. The Pascagoula health department in Jackson County shared a small building with the health department in neighboring Ocean Springs and also set up shop in a tent in the parking lot of their former facilities in Pascagoula.

Stokes wanted to help. But because the foundation's emergency fund was not flush, he first issued a call for support. "Kaiser Permanente made a lead gift of $2 million to the Emergency Response Fund, and the Robert Wood Johnson Foundation contributed $1 million. Hundreds of other individuals and organizations gave generously. These gifts to the Fund enabled the CDC Foundation to immediately respond to requests for help."

As Travnicek had instructed in early September, Stokes talked with District Administrator Kathy Beam about the best way to move forward. She had assigned Michael Cruthird to be ESF-8 liaison between Harrison EOC and the Department of Health's regional command staff. As such, Cruthird had developed a comprehensive plan for restoring primary health care services and proposed using modular units for facilities in Hancock and Jackson Counties while construction of permanent facilities could be completed. The Central Office did not act directly on Cruthird's proposal, but Central Office Administration Director Mitch Adcock submitted the official proposal for both buildings and quickly turned the project over to Kathy Beam.

Stokes and Beam worked to replace tents with the modules. Within months, CDC Foundation helped MSDH acquire two specially designed modular public health clinics to replace destroyed facilities and restore public health services in Hancock and Jackson Counties. Part of the foundation's process was to check with the CDC Emergency Operations Center on any request; CDC's EOC confirmed that the project would be a critical part in helping to rebuild the public health infrastructure and get services back in place.

"General Electric made those buildings and gave us a good deal on them, $500,000 each, as I recall," Stokes said, "and I sent them a check for a million dollars. A couple of weeks later, the CDC Foundation held a regular meeting of its Corporate Roundtable on Global Health Threats, on which sat one of GE's medical directors. He said that GE had a plastics plant in Bay Saint Louis and that the health department was surely providing services to GE's displaced employees and that I should return to GE and ask them to donate one of the buildings. So I gingerly picked up the phone and asked if they would donate. A week later, they wrote us a check for half a million dollars; they said to use it to help the local health departments and CDC respond even more."

Beam acknowledged the foundation's gift. "In the midst of all the trauma, the assistance from the CDC Foundation fund was like a life line." Travnicek would credit Former State Health Officer Ed Thompson for "all the heavy lifting on all these buildings we've got. Thompson had told the CDC Foundation people we were destroyed. That's how the CDC Foundation became the one group of people who gave us exactly what we needed."

Installation of the modules as a temporary fix gave Travnicek and his staff time to work through county and state logistics toward planning, funding, building, and opening state-of-the-art, permanent public health facilities in Hancock and Jackson Counties. At the Jackson County dedication in 2010, he said, "I knew we had to buy time; I didn't know we'd have to buy five years. So Charlie bought time for us. We'd hear about roadways to nowhere; this is not a roadway to nowhere. The bricks and mortar here—these are ideas that happened."

§

Significant in its own right and for specifically benefitting public health, and therefore the entire population, the foundation's gift joined similar largesse from every state and many nations. Within weeks and over time, individuals, foundations, and corporations donated billions in cash and products; others volunteered more than 14 million hours to clean, repair, or rebuild homes, serve meals, and perform community service.

"People were wonderful," General Spraggins said. Starting with people who lived and worked in the local community, "They just came in to help. Chancery Clerk John McAdams came in and said, 'I'll do whatever you need me to do.' He wound up taking care of all my distribution points. Everybody and his brother called, wanting to come to South Mississippi to help us. They were calling from France, Austria, all over the world wanting to come help us. Thank God for people who were doing that! Lots of churches, cities—they knew we'd lost everything we had. They were coming down, and they started bringing everything we could have wished for.

"We got a warehouse on 34th Street and some big tents, too. Of our main distribution points, one was at the Coliseum and one at Long Beach. They would set up tents to distribute food; people could go in with a box and get what they needed and come back out. Well, they set up clothing tents, too. We talked about it, and one of the things I wanted more than anything—when a person has lost everything he has, you don't want him to lose dignity, too. John McAdams got with Goodwill and they gave racks; so volunteers were putting the clothes on the racks. And then we were going to have people to work the centers and organize and fold the clothes—for example, if a person needed a medium shirt, the people who worked the centers made order, conveniences for the people who had needs, so that in going there, finding your size, getting what you needed—it all had dignity to it, not like picking up something off a trash pile. And they had good clothes—a lot of them were brand new.

"Bill Holmes let us use the Convention Center/Coliseum to house and use as a central distribution center. We had a system, and it was working, and people had dignity; that was great about it. It worked fantastic. Finally, people had what they needed. Goodwill took the rest of the clothes and sent them to be used all over the world. By the grace of God, it didn't rain before we got 98 percent up to be used in here or in some other part of the world. Not a dime of anything, not one shirt, nothing went to waste."

§

Myrtis Franke stands among Gulfport residents who also lost her home and all its belongings but kept on working and trying to assure that South Mississippians regained their standing to recover and rebuild. Having evacuated pre-Katrina to her daughter's in Jackson, she commuted daily for two weeks and then moved into a rent-to-own property. Then on the Gulf Coast staff for Senator Trent Lott and chair of the Memorial Hospital of Gulfport Board of Trustees, she pulled every string she could to help.

"At Memorial Hospital, people were crawling in the damn storm," she said. "I had called and talked to General Spraggins. People were begging us to come get them off roofs and out of attics. And after, Memorial didn't know day-to-day if they were going to have fuel. They cried together when things were bad, and they laughed. I cannot imagine what they all went through, not knowing if their families were okay. We were working long days, seven days a week. You don't think about it because there's so much work to be done. So much hurt."

Declaring firmly that Memorial stories could fill a whole book, she talked about The Village. "They had to set it up and let people stay there; they had no home to go back to. So people actually lived there. They fed them, got cots for them, and they lived there. And they went from seeing six to seven thousand people a year to several hundred a day. Doctors came back and volunteered, helped to triage. There were cuts, people working with equipment they're not used to, and there are nails everywhere, glass everywhere—and no shoes."

Even though she personally knows "you can't look back" and tries to live that mantra, Franke feels for people who suffer. "All the congressional delegation were down here. It took Trent at least a year to get over the loss of his home; he struggled with it—if you ever get over it! He was very emotional about it. I found him one day when I came in—I flew one day with the National Guard over to Pascagoula to look for him. And we found him a few blocks from where his house had been, he and Trisha looking for things. We got into the helicopter and toured the whole area; it was amazing. The only way to describe it is to say you couldn't believe what you were looking at."

Formerly on the State Board of Health, Franke knew the need for tetanus shots. "And the state health officer, who's no longer the state health officer," she told the story in 2010, "had not even been down here when I had a discussion with him on the phone one day. AMR wanted to get out there and give shots, and the health department was going to give them a few vials of vaccine; so I drove up there and talked to them. I said, 'No need to give them a few—if you give them a thousand, they're going to be right back; give them enough.' The director told me that we didn't need a thousand shots down here, and I told him, 'Well, you'd better come down here and take a look.'

"Experience," she said. "That's how you learn to do things better. Dr. Travnicek was awesome. There was some division between the department and the EOC, but he kept doing what he knew to do and what he needed to do. We got involved with that. . . I had a good relationship with the director and tried hard to get along with everybody. And got him to leave Travnicek alone and let him work. Because General Spraggins needed him.

"People rose to do whatever needed to be done—incredible! And they've done it from one end of the Coast to the other. People worked—I know people who had water up to their ceilings, and people jumped right in, pulling their own sheet rock. People have gone through some incredible times! They're still putting their lives back together, and they're still smiling."

Not everybody, however, got the immediate attention and help they needed, she acknowledged. In late October, despite having been assured that all acute needs had been met and that nobody remained there, Franke drove to Pearlington, a community of about 1,300 west of Waveland and on the Mississippi-Louisiana border. "But they were there. They were all in the woods. All on slabs or little tiny tents. Whole families, and it was emotional. If they got a trailer, they didn't have a pillow, they didn't have sheets, they didn't have blankets. Red Cross wasn't set up there yet to feed them—oh, many horror stories. In my opinion, some folks in Hancock County suffered the most. I had friends who lost everything. Their homes, their cars, everything. Give them canned goods, and they didn't have any way to open it. It's hard to imagine Hancock County's resilience, but they never complained. Never, even though it's been hard to find everybody out there who needed help and try to help them."

When Red Cross did arrive to give out "cold lunches in an area so devastated, they had no way to fix hot meals. And when I'm telling you— see this desk? There were whole families living in a tent that size. We found a Vietnamese family living on their boat way up in the woods in Hancock County; they had enough provisions to last until they were found. A woman called about a body in a tree in her back yard—nobody knew it was there. People have incredible stories, and their strength and attitude is absolutely wonderful. Multiple people living in a school that had been terribly damaged and water all in it when they were in there—and you could smell the smell; you know when water's been standing—but people slept in it. They had no other option. Where were they going to go?

"We've had people forgotten in tiny trailers," she described. "They had to get out, to move. One woman with five children—we had to get her out of that thing because they're not made to live in 24 hours a day.

"We had somebody 83 years old in Hancock County living in a lean-to, a shed, and ran a cord from the neighbor's to get electricity in there. No bathroom, no kitchen. It's astonishing what people had to do, and they were not knocking on our doors begging for help. They didn't know where to go for help. The people down here—the attitude has just been phenomenal. We're making it work. We're helping them do this."

And she heartily acknowledged, "We couldn't have done any of this without the volunteers. People have been incredible; volunteers from all over this country, in enormous numbers. And they're still coming. Church groups and the Menonites; they came and worked and built houses. People have lived in less than optimum conditions themselves in order to do this— that's what's amazing to me. People have come from all over, and they can't even get a bath easily. They've done it with no complaining, and they come back. We still have people with damaged roofs that have leaks. We're in a tougher time now—finding the tougher ones to help. And some that are afraid to get out of where they are think if they stay there, it'll work its way out. One woman came from Texas and another from up North to volunteer, and they ended up going home and moving here to help."

§

Sherry Lea Bloodworth, the Fairhope, Alabama, woman who helped evacuate nearly a thousand special needs shelter victims just days after Katrina, also moved to Mississippi about 15 months out. She transitioned from church-affiliated, to independent, to Angel Flight volunteer, and, finally, to working with Architecture for Humanity.

"I just landed in Bay Saint Louis," she said. "People were still sleeping on paddocks so I helped get a shelter started at 2nd Street Elementary with 300 beds. The Red Cross would man the shelters, but there was no security, no power, and no potable water. So I had to help put together partners who would come in to bring food. The Baptists, the Scientologists, the Jewish community—everyone came together, and then the youth teams ministry came and set up a kitchen. I got a call from a woman based in Los Angeles who was connected with the Jewish network through Jackson; they called from Jackson and offered to bring me supplies. It was just a matter of filtering resources. That's when Architecture for Humanity found me and hooked up, and I knew, based on their work and what they do—they work on post-disaster housing issues—I knew I should be working to help provide housing."

A get-things-done individual, Bloodworth intercepted Governor Barbour on one of his many trips to the Coast. "They said I had two minutes. I said to him, 'I want you to accept their help.' And his chief legal officer came and told me they would come to me in Alabama, or I could come to Jackson—however I wanted to do it, 'but we want to talk to you.' So I went up to Jackson, and they put me on ESF-14's transitional housing committee. And I started working housing issues and have been involved ever since."

Through Architecture for Humanity, she helped get a four million dollar grant from Oprah's Foundation, The Angel Network, to start a couple programs in Biloxi. She started two different funding pools, one to address rehabilitation and help get the resource center up and running and another to build seven varied-elevation model homes that meet the building criteria for disabled families, for people with sick children and other challenges.

"Everything has to do with housing," she said, "advocacy, partners, Mississippi cottages, the formaldehyde stuff, how to get people quickly into an alternative form of housing—we have to present solutions!"

§

Gulfport's own George Schloegel stands near the summit among Katrina responders who became heroes for helping people re-start their lives. His own history with hurricanes helped influenced his aggressive action in 2005. As a little boy in 1947, he endured an unnamed hurricane that "came and swept east to west down the coastline." Several minor hurricanes occurred before Betsy in 1965, a "pretty tough storm that broke one of the levees in east New Orleans, and the whole area flooded, but New Orleans did not learn. In 1969, Camille taught us a lot, but Katrina taught us more."

Schloegel compared the two monster hurricanes that catastrophically destroyed the Coast 40 years apart: "Camille was a stronger storm but compact—small in size and much more powerful than Katrina. They estimate winds probably got up to 227 miles an hour for Camille. Camille came in and hit at Pass Christian, right across from Bay Saint Louis. She was in and out in a matter of a few hours because it was small and fast-moving. It was worse if you were right there in that compact period but since it didn't last so long, it wasn't as bad as Katrina. Katrina was from Texas to Florida and covered almost the entire Gulf of Mexico, and it took something like 14 hours just to pass through. So, while the winds were lighter, Katrina was pushing that water longer, the water got higher than in Camille. Let's put it this way, you might could take winds at 227 miles an hour for 15 minutes; it's a lot worse to take 175 miles an hour for 14 hours! That was the difference. Duration and size."

Prone to look for lessons learned, Schloegel affirmed, "After a calamity of any kind, the normal operation of commerce is abolished."

When Katrina slammed ashore in South Mississippi, Schloegel was, among other positions he held for Hancock Holding Company, president and chief executive officer of Hancock Bank, established in 1890. "The things you normally do—like writing checks, going to ATMs, getting groceries—all of that is out. Number one, many places are not open; number two, electronics go away, and we're dependent."

Schloegel's family and 150-year-old home safely survived the storm, but damages at 59 of Hancock Bank's 155 properties in four states totaled about $49 million; 22 sites suffered more than 50 percent damage. About 10 percent—190—Hancock associates lost their homes, and more than 400 and their immediate families suffered significant property damage or loss. "Our first priority was to get a bank open! We opened the bank the next day, even though we also were absolutely wiped out. We opened in the areas where we were not damaged, but our footprint was the footprint of Katrina so almost all of our places were damaged. Our objective at that time was to get cash to the public. No ATM would work. People did not have the wherewithal to come into the bank. They may need only one, two, three hundred dollars, or they might need ten bucks to buy some shorts or a few dollars to buy ice.

"People who were working in the bank were whoever could get down there in their flip-flops—and we had hundreds who did that. But somebody who hadn't been in a teller's cage for 30 years all of a sudden had to go back to being a teller. We were able to retrieve some of the money that was in some of the vaults where the bank had been destroyed or in an ATM. So we set up a battery of washing machines and dryers in a building where we could get a generator to generate our own power, and we laundered money. Literally. Washed it—we had to wash all that mud off it. Our people were laundering money and ironing it," he chuckled about a national publication's photograph of the operation with the caption "Bank Launders Money!"

Having worked at the bank since 1956, he knew that would not be enough. "The economy needs a lot of money quick." Even making a telephone call took extra trouble, but bank officials were able to contact the Federal Reserve in Atlanta and one of their own young just-out-of-college associates who had evacuated the Coast to his parents' home in Atlanta. His instruction was simple: get into your Suburban, go to this address, and bring back to Gulfport as fast as you can what they give you at that address. "What they gave him was $30 million in cash, and that youngster drove that cash here, and that's the cash we used. We also doled that money out to the other banks." Bank examiners disagreed with some of their "innovative thinking," but Hancock Bank served its community.

The bank had an operations center just across the street from Harrison County Courthouse. That's where the money laundering happened and where staff met twice a day, at 7 am and 3 pm, with whoever could get there. Personnel's duties changed, putting some into jobs they had not done in years and opening doors to totally new work for others. One might go get gasoline for the next day; executive vice presidents made sandwiches and went for water. "We made a decision that anyone who came to our bank, whether they were Hancock customers or not, we would serve them, and we would give them at least $200 in cash. Now I want you to imagine. You live here on the Coast, and you bank with some bank that wasn't open—never been to a Hancock Bank before. Your house is washed away, and all you have is some of the clothes that you have on right now. You don't have identification; you don't have any check; and the word is out, 'Go to Hancock Bank, and they'll help you get back on your feet.'"

Accepting IOUs and a signature, "We gave away over $3 million to noncustomers who put that piece of paper in a box. The bank examiners went ballistic," Schloegel recalled. "They said, 'How in the hell do you think you're going to get that money back?' We said, 'We won't get it all back, but we'll get most of it back.' And so, with these old-time IOUs in that box, we lost less than $300,000. . . and in the five months following the storm, we got 13,000 new accounts. The bank grew in total deposits as much in those five months as we had grown in the first 90 years of business. . . In the worst conditions, you take care of your people."

Without doubt, Schloegel believes, "The very best comes out in times of crisis—that's where character really shows. And you could see that all over this area. The outpouring of assistance to the area that was hurt was so incredible. The Baptist Men's organization out of North Carolina came in, and they didn't walk in empty-handed. They came in Winnebagos and trailers fully supported with a lot of food and a lot of chainsaws. . . so you didn't have to take care of them; they took care of you. Right next door here is Gulfport Fire Department. The Fairfax, Virginia, fire department came here with relief supplies and set up and stayed for a couple months. College students on their Christmas break and spring break, instead of going to Florida to have fun on the beach came here to help the old people clean out and rebuild their houses—and they're still coming, in 2014! Don't ask me to explain it. I can't explain it, but I can tell you that whatever the force is called—God or whatever you want to call Him—the lessons of a calamity are part of what makes us what we are. There is more to be learned from the hard times than from the easy times."

Three of Schloegel's own children lost their homes in the storm. Grandchildren who were youngsters then would now say they "learned what's really important in life. Most of it's just stuff. It builds character, and the character that comes out makes you feel good about humanity. . . Now, I don't want another storm; no one does. But when it comes, I can tell you we will deal with it and be better off when it's over because of the character-building that takes place on one neighbor's helping another neighbor."

Chapter 30

Unbelievable destruction, incredible response from victims and volunteers, determination to persevere—champions of Katrina in Mississippi lived through it, survived it, and incorporated their new-found strength and resolve into rebuilding their homeland.

Spraggins: Katrina was an act of nature; what we do after her is an act of God.

Weaver: It will happen again. We're in the tropical path. Regardless of what the disaster is, the best and worst of humans will present. Hopefully, the best of the people and our response will be remembered.

Rocko: I think we'll be better prepared. Camille killed more people in Katrina than Katrina did because they used Camille as a benchmark. My goal from all this? To help make life better.

Lacy: Our success stopped selling news. Our story was not blood and gore. Show me people on the roof of the house and let these people be pissed—they're more of a story. 'Our cup runneth over!' That's just the news media. The Harrison County story: we came together, worked as a team. We had feelings, irritations, but you have to overlook them sometimes. It's about the citizens, redeveloping the infrastructure.

Hargrove: I keep thinking about what we do know and what we have not yet learned.

Delahousey: Look at what Mississippi did, and we did some stuff right: Department of Public Safety, public health, our governor. It's a good thing to have a Democrat Congressman and Republican Senator who both lost everything and are working together. Steve Holland and Alan Nunnelee get along after hours, and they do anything they can to help us. Haley Barbour says no other state can say they didn't lose anybody in a hospital or a nursing home—and next time we'll do better!

Stokes: While most people understand the role of first responders—firefighters, police, and emergency medical personnel—the work of public health experts before, during, and after a crisis might be less known but is always equally important.

Patterson: I hold the belief, and I think it's supported by research, that people's memories are brittle. We lose sight very quickly of how bad "bad" can be. It's very timely if you can communicate through the stories not just how bad "bad" can be but the extraordinary, sometimes extreme, measures necessary to overcome catastrophic disasters.

Travnicek: Was it one bold move after another? Yes. Was it risky? Yes, but I was willing and, the thing is, it was moving so fast. . .You have to have the heart of a lion—most public health workers have never seen anybody who's practiced medicine with the heart of a lion, and that strength of purpose can be very scary to the uninitiated. We are diagnostic dilemma people. This is setting up relationships and knowing in the heat of battle who will run and who will fight. And you have to decide what ditch you're going to die in, and I decided that ditch down here was the ditch I was going to die in professionally. I was prepared to die in that ditch.

§

Chip Patterson led the Jacksonville, Florida, Incident Management Team that helped Harrison County EOC develop its unified command structure; in 2005, he was named Florida's Emergency Manager of the Year.

"One of the things I've seen over the years," he said, "is most times, a community's leadership without disaster experience—particularly absent a lack of imagination or knowledge of how bad 'bad' can be—also lack imagination regarding the extreme measures a community must take to overcome this disaster, to loop from chaos back to recovery. Communities are staffed, equipped, budgeted for normal day-to-day issues: social, infrastructure, normal day-to-day problems. Those are challenging enough. And then disaster comes and requires a different type of thinking to deal with disaster—requires a different type of organization and intergovernmental relationships that do not occur day to day between local government and the state and the federal government. Those are some extraordinary obstacles to overcome.

"And most people point to FEMA: FEMA will fix that, FEMA will solve that, FEMA will overcome those huge barriers. There is no question that FEMA does represent the government and has a huge part in problem-solving in the face of a major disaster. But it truly starts locally; it starts in that community, because they know what the critical facilities are, the critical infrastructure, what it takes to be able to provide and to thrive through that disaster."

When the veteran emergency preparedness and response manager brought his team to Mississippi, they came not only with vivid imagination but also fresh knowledge of how bad "bad" can be. In 2004, Florida lurched and whirled at the brunt of four hurricanes. Florida Division of Emergency Management commemorates "the record-setting 2004 Atlantic Hurricane Season [that] brought Floridians four hurricanes in just 44 days. Hurricanes Charley, Frances, Ivan, and Jeanne will forever live in the minds of Florida's residents and visitors as some of the most destructive and dangerous storms to ever hit the United States."

Florida emergency responders started watching Katrina before she gained hurricane strength; they watched her emerge as a disturbance near the Bahamas and experienced death and destruction after she lumbered ashore near Hollywood, Florida, three days before she would blast Mississippi. They knew that wherever else she might land, Katrina's consequences would devastate.

Governor Haley Barbour in Mississippi and Governor Jeb Bush in Florida also tracked the huge storm. Bush pledged support to Mississippi.

By October 6, 2005, Florida would commit aid valued at just over $128 million to Mississippi alone. More than 6,000 civilians, National Guard, and EMAC response teams would deploy to assist in Mississippi. Florida sent rescue units, tankers, command vehicles, water search vessels, and communications units. From Florida Department of Health, 384 medical assistance personnel delivered advanced life support services, vaccines, oxygen tanks, and logistical support officials. Hospitals in Tampa and Miami accepted 86 patients transferred from Mississippi. Florida sent 954 trucks of water, 940 trucks of ice, 114 trucks of MREs, 8,438 cases of baby food, more than 10,000 cases of baby formula, 4,000 cases of Pediasure, 2,100 cases of Ensure, 16,000 cases of juices, 1,755 cases of diapers, and 2,495 cases of bottle nipples—among other relief supplies and personnel.

When Patterson's team reached Gulfport from Jacksonville via the Florida Emergency Management Agency in Tallahassee, they approached with what he calls "respectful aggression." They advanced, he said, "respectful of jurisdictional responsibilities, authorities, needs, but at the same time aggressive in meeting identified needs for the community, being willing to provide all that's needed, not just immediate needs. . . In a major disaster, the provision of bottled water and ice is not that simple. It's not as simple as load a truck, take it into the disaster area, and drop it off. . . It requires an organizational structure, logistical support, communications, and coordination necessary for that bottled water and ice to be delivered effectively to the disaster survivors, to meet the need. Much of that is 'lessons learned' from Hurricane 2004 season: that coordination, communications, and infrastructure must be in place to effect the necessary logistics."

Teams similar to Patterson's came to Jackson and Hancock Counties. All operated under the umbrella of the coordinating officials at Florida EMA. They brought from Florida everything they would need to be self-sufficient. Patterson's deputy, Martin Senterfitt, remembers that "when we pulled out of Jacksonville, we actually grabbed a 22-foot travel-trailer to sleep six, and we loaded it up with and took our own food. We had a team of about 18 people and were sleeping 14 people in this six-person travel-trailer—we were piled up like puppies! By the second or third night, we just collapsed; everybody would lie down wherever we could and were averaging about four hours sleep a night. Every morning when we would wake up, we would walk outside and see many, many residents of Harrison County were sleeping on nothing more than garbage bags to keep water from seeping through the ground. Even though we were packed, at least we had a roof over our head. Most people didn't even have that."

Some of the teams from other locales brought no resources. "All they did was come in and give a lot of orders," Senterfitt said. "We would have these teams sweep through the county and try to change everything and then move on to someplace else. But what they didn't realize was that Dr. Travnicek had already organized, and all they were doing was messing us up. We did a lot of talking about moving shelters. We did deep thinking about how we would handle the problem, and the next day, before we know it, another health department from another part of the world came in and started trying to change everything. And when we would challenge these other teams, the first excuse they always use was, 'Well, there is no government, it's all chaos; so we're here to bring order.'

"And we said, 'No. Number one, Harrison County, Mississippi, never lost control.' Contrary to popular belief, Harrison County never went into anarchy. They were organized the whole time; they were just not back on their feet yet, like anybody who just took a 40-foot storm surge would be. But Dr. Travnicek had it organized. The only time that the Health Department truly experienced problems was when outsiders came in without reporting in to the county emergency management system.

"When Dr. Travnicek would push back . . . It was people from Mississippi; it was people from Florida; it was people from the national level. And Dr. Travnicek every day—five, six, ten times a day—would look at them: 'We have a plan. Follow the plan. Quit getting out of line.' They would push back: 'You're out of control.' No, we're not; you're causing the problem. Settle down; we're working through this.

"Just his level of leadership!" Senterfitt remarked. "Dr. Travnicek is a little bit quirky. He's got kind of a personality. But he was in his element and was a very calming influence. This man knew what had to be done; he was the most qualified of anyone to understand what we were dealing with."

Travnicek knew that he knew what he was doing. Going into the emergency operations center before Katrina's landfall and being there to participate in response meetings—that's what he expected of himself. As county health officer, he had statutory authority and responsibility to protect the public health. From previous storms, he knew that coordinating services and resources for hospitals and nursing homes would require immediate before-the-storm attention, and he knew that concerns after landfall would fall largely into the environmental health area. That's when the focus would be on food, water, sanitation, and safety.

Inside the EOC and a vital member of the response team, Travnicek did public health and took care of fellows in that command center. "I was their doctor," he said. "And whether they were upset, got cut, needed a shot, or what the hell they needed, I'm their doctor. They need sewage? I sign off on that. I sign off on the chickens getting removed, sign off on pork bellies getting removed, go out, see these people. I drove up and down looking at pork bellies and stuff like that. They would drive me around, and I would say, 'Okay, we're going to condemn all this.' I condemned millions, hundreds of millions of dollars worth of stuff. One time I condemned three million pounds of seafood, and they came back a day later and they said, 'Geez, you know, we went to the thing, we bulldozed, we took . . . if we take the shrimp out, we're going to have to bulldoze the whole thing to the ground. It's a six million dollar loss,' or something like that, and I said, 'Okay, where do I sign?' They said, 'Sign here.' I signed it. Hell, within 12 hours it was bulldozed to the ground. It was all swirling around me. If you can handle quite a lot of things, and you're willing to make immediate decisions and not dither, then you just go do the things. A lot of things were happening."

Public health issues comprised part of almost every day's sit rep meeting. Patterson recalled one problem "related to the hospital that seemed to change the direction and the tone of what we were doing to support Mississippi. Every evening at six o'clock, I had a briefing to provide our area command in the State EOC for Florida in Tallahassee—our state emergency management director then was Craig Fugate, now FEMA director. We had just received reports at Harrison County EOC and had started to understand the profound problems at hospitals with water, with not having clean water. And I'd just gotten a report that infants at the hospital were dehydrated. There really was no indication at that point that we could improve that, that we could get clean water for formula or for hygiene at that point, and they were going to have to abandon the hospital and transport with all the attendant problems of moving critically medically frail people that evening. I did not realize that the state director had the Governor of Florida sitting beside him—I made the statement that if we did not do something very aggressive, very profound, there would be more loss of life on the Gulf Coast. To their credit, the Governor of Florida, Jeb Bush, and state director connected with the Governor of Mississippi and FEMA and worked out the details for resourcing that resulted in those numbers about dollars spent and people sent."

Only when Harrison County began to see the trucks arrive and people come in could they begin to grasp what Florida had done toward saving lives and assuring, at least, short-term recovery. Similar response poured into Hancock and Jackson Counties. As the Florida team worked with their Mississippi counterparts, they also assessed both the responders and their work.

"General Spraggins was brand new in his job," Patterson said. "As a result, he didn't know all the lingo that has a lot to do with first responders, hospitals, and public health. He did not necessarily know relationships in the county and state—though as a National Guard general, he probably had a good idea—and certainly not the relationships with FEMA, but I observed that he knew how to be an executive leader. He had to do things that required courage and capacity. He had to do like many other people in an environment with no real time to reflect on consequences about decisions and still do the tough things, still face facts, even though they're facts we don't like. And he did all that. I was able to very closely observe him, both things he publicly did and what he said privately. I don't know anything about his previous career—only that brief 10-day stand that I'm referring to. As credit to Harrison County, you would hope that every elected leader would identify their top folks, and I think their current emergency manager is the same."

Patterson turned his attention to Mike Beeman, "one of those rare people in FEMA. FEMA has a host of people like Mike Beeman that FEMA puts on the point of the spear—I guess that's a good way of putting it. Mike was sent to Harrison County to represent FEMA and support the county and to be able to communicate back to FEMA at large what's occurring. He went above and beyond in his service to the county, with a heart for service, that heart for doing good things in the middle of a disaster. His information and insight proved invaluable to General Spraggins, to County Administrator Pam Ulrich, and to the Board of Supervisors, because their responsibility in the middle of all this is doing things to save lives, protect property, and restore community, but also being diligent to assure resources—expenditures—are used to align efforts with FEMA reimbursement guidelines. Mike did a very good job and provided sound advice."

That "heart for service," he suggested, comprises the core of "most people working in public health. Most people who work hard in public service are not there for the glory, the recognition, and certainly not the pay." Turning his focus to Travnicek, Patterson continued with a bit of humor and all seriousness: "Bob is a get-it-done guy. In his job, his role, he was a big-picture guy. He understood all the public health problems and issues, and that's necessary in that environment, but as critical, or more so, is that he is also a detailed guy who works to solve problems. Marrying the two of those—being able to think strategically and act tactically in the midst of all sorts of bad things going on—is difficult. To think strategically is pretty difficult sometimes anyway, and within the complexity of bad things going on, to break it down and act tactically in a disaster environment is a necessary thing. Few people are able or willing or have the courage to fully utilize emergency authority that most states and local governments have to deal with in a catastrophic disaster.

"Those authorities are different for extraordinary circumstances compared to what we all work on during our day-to-day lives. We have laws, constraints, and regulations that are necessary but that in disaster time, there are mechanisms or waivers for those laws and regulations by an elected body or elected leader. It's one issue to have them waived by a governor's executive order or a local designation by a Board of Supervisors; it's another matter to act on it. Dr. Travnicek was willing to act in extraordinary times," said Patterson.

In that regard, Senterfitt likened Travnicek to Phil Connors, the protagonist (played by Bill Murray) of the film *Groundhog Day*, particularly in his ability to break the loop, to break the cycle. That capacity, he said, based on the Katrina experience, became a requirement for every community's disaster preparedness plan.

"Before Katrina," he explained, "under the old national model, we put people into shelters, and after the storm had passed, they were going to leave the shelters. The problem was that in Hancock and Harrison Counties, they had no where to go; their homes were destroyed. So what would happen is people would stay in the shelter, and every day they would be back in the shelter, and we got *Groundhog Day*. That's when Dr. Travnicek and his people started talking about 'we have to break the cycle.' As long as the people were cooped up—several thousand people—in a shelter without basic commodities, we're going to wake up every morning and face the same problem, and it's only going to get worse. So Dr. Travnicek, working with some people, actually came up with a plan to evacuate and to move people outside the area, to places that do have water, electricity, and the ability to shower. They set up a whole plan of how to evacuate these shelters, to get people to an area that can support them outside the disaster area. Sheltering in and of itself is not enough. You've got to have a plan of what to do after the disaster. Now, post-Katrina, that's become a required part of the system—it's not good enough to have a plan on where to shelter but also what to do with people after the storm passes.

"I know Dr. Travnicek was challenged on multiple occasions," Senterfitt continued, "and he would always finally smile and—with that quirky grin of his—he would say, 'Yes, but here's how it is and how we're going to have to do it.' You know, Dr. Travnicek likes to talk. And he would just sit down and the longer he would get pushback . . . He would talk them into submission; they would say, 'I'm going to have to agree, or this man's going to keep talking to me.' They would give in, and he'd get his way. He was never nasty; he was never rude. But he was so knowledgeable that people that would come in and try to push him around—he'd give them the talk. And he could talk longer than they could tolerate, and he'd just win his battle and go on to the next. It's people like that who really make a difference in times of disaster."

Not everybody appreciated, or even wanted, Travnicek at the command center. Greg Doyle, who often represented Delahousey in AMR's ESF-8 seat beside the health officer, knew that Brian Amy apparently wanted Travnicek out. "Travnicek was profoundly important," Doyle countered.

"He did need to be there. He was calming and calm and caring. He played a huge role in facilitating information, and he caught a lot of grief. He's the health officer over six counties and I remember him dealing with issues over the six coast counties: water samples, inspecting nursing homes, getting restaurants open—he was available to everybody."

With both AMR and the county health officer working from the same desk, they directed ambulances to do double-duty, acting not only as first responders but also in delivering printed public health information pieces into the communities and bringing situation assessments back into the EOC.

"People of Mississippi were knocked to our feet," Doyle said. "The difference here is we said, 'Help me to my knees.' Others said, 'Get me to my feet and give me stuff.' Cohesiveness of the communities and leadership in Mississippi made the difference."

§

"I do want to tell one revealing story of the impact on people and first responders," Patterson offered. "Whether it was Tuesday or Wednesday after the Monday storm, one of the things we knew to do—you certainly take action on what is immediately observed that a catastrophic storm has done. Just as quickly, we needed to get some real assessments accomplished. In this case, a public safety assessment of fire and EMS capability in the southern end of the county. We had two fire chiefs, both from fire/rescue agencies from Florida, there to augment the whole assessment mission, to help the county assess. We were working to coordinate through the county fire coordinator. We had a meeting with the fire coordinator, had the two fire chiefs from Florida, and the county coordinator. I had already cleared it through and advised General Spraggins and Rupert Lacy that we were going about this. Our request was that the county fire coordinator go with them to the southern end of the county so he and they would see the same thing in a factual, data-oriented fashion. And he said, 'No.' He refused to go. I was surprised. I thought it was a transportation issue or curfew—the exclusion zones were very stringent—and I said, 'These guys can get past the checkpoint, they've got the credentials, the right kind of marked vehicles; go with them.' And he said, 'No, I'm not going to do that.' He refused again. I got kinda upset with him—is this some sort of political thing? A county guy who won't be involved in public safety assessments? He's just going to serve firefighters in the northern part of the county? That's what I was thinking and I was upset when he said, 'No, I'm not going to do that.' . . . I expressed being upset with him and said, 'You've got to do this; this has got to be accomplished'—because I'm thinking ahead to the six o'clock report and request for resources meeting. And the guy gets emotional with me and says, 'You don't understand. My family lived in Pass Christian during Hurricane Camille, and I lost both of my parents in Hurricane Camille. Most of us moved up county after that storm, and I still have cousins throughout that flood zone I have not heard from. I cannot go. I cannot go.'

"I was floored. I felt about three inches tall. As Chip tried to resurrect some honor, or some dignity, I let him know these fellows would go and do the assessment and report back to him so he could put his stamp on it or reveal it on his own behalf. . . I think there were probably dozens and dozens, possibly hundreds, of people working, doing their job for the community, and not knowing. Whether they were working in government, working in a hospital, or working in a safe hut with a volunteer organization —I think there were hundreds of people who did not know the fate of their homes and loved ones but just kept working."

§

As lead FEMA representative on the ground for the first eight hours of the 9/11 World Trade Center disaster, Mike Beeman looks for "lessons learned." Comparing, he suggested the New York City event produced a "very small footprint. Katrina's is larger than the whole landmass of Great Britain. Katrina looked like a war zone." A young pregnant woman taught one lesson that he remembers and shares.

"The story goes that a young lady about seven months along had climbed into an attic in Harrison County and had her cell phone with her. After she climbed up there, she realized she could not get down and spent a number of days in the attic. She had dialed that cell phone, and when she got the 'call failed' signal, she cut the phone off to maintain power on the phone. *A very smart lady.* Somewhere after—I heard it was seven to eleven days—she connected with somebody in Indiana or Illinois who connected her back to the EOC, and they got emergency responders to her and got her out of that attic. She was smart enough to not continuously dial the cell phone and to save her energy.

"One other thing we did in the EOC that was extremely beneficial— and I now preach this everywhere I go—if you want to learn something valuable from Katrina, it would be involvement of the private sector in your emergency operations center. The most critical elements of the private sector you want to have in there are a senior representative from the telephone and cellular companies. They'll be the ones who help you prioritize the restoration process. Every day, Bob Fairbank would come up to me at the table after those meetings and look at Bobby and Rupert and me and say, 'Let me understand what I think should be the priority of the day based on what was discussed in this room. You guys said we're out of fuel at Hospital X'—I remember that happening one day. And we said yes. 'So I'm going to shift my crews to get power restoration back to that hospital; I'll make that my priority.'

"We had representatives from Florida, representatives from the area that had been hit by one of the storms two years before. They came in and drove around and said, 'You guys are seven weeks ahead of us—how did you do this with power restoration and so many other things?' And everybody pointed and said we used the structure. We used the incident command structure which gave us a concept of operations that set us in the right direction, and we never wavered. Harrison County was one of the first to use ICS in such a catastrophic event."

§

George Schloegel took on responsibilities beyond banking. "We were right across the street from the Courthouse and the EOC. Right after a storm because of the debris, the heat, dead animals, and so forth, public health becomes a real issue. And you've got boards with nails in them and the possibility of tetanus, typhoid, and other diseases. You have to quickly get an immunization program going to make sure those people who are working in that kind of debris get their tetanus shots. How do you administer tetanus shots when it's all you can do to find a drink of water and have no transportation? Travnicek, and I don't know how he did it, but he ended up with a truckload of vials to give shots to people. The word was, 'Go to hospitals to get a shot.' Doctor's offices weren't open; drug stores weren't, and getting to the hospital is almost impossible. We needed to open a bank and didn't have time to walk five miles to the hospital and come back to work; so we said we'll take care of that. I went to Travnicek and told him, 'We've got a problem; a lot of people are coming from out of town to do all these things. I need some tetanus serum.' 'How much do you need?' he asked. 'How much can you give me?' 'Would a case of 600 help you?' 'Yes!' He gave me a box of 600 and said, 'Can you handle it?' Well, we raise horses back here, my son works at the bank, and we do our own vet work; we give our horses shots. So we lined those folks up over there and he'd pop them. 'Have you ever done this before?' He didn't tell them he did it to horses—and we inoculated 600 people. My son knew to use a little alcohol and clean 'em up when I handed him that vaccine. Was that the right thing to do? Was our public health officer sticking his neck out? Yeah! But you don't ask for permission; you ask for forgiveness after it's over. You have to do the right thing. Serving the public and whatever the needs happen to be is what you do, and that's what Travnicek did. He made sure that the health of our people was first and foremost in whatever he did even if he had to cut a corner to make it happen. He's a no-baloney type of guy. In a crisis you want him right there with you. I can say exactly the same thing for Rupert Lacy.

"Are you familiar with the management book *First, Break All the Rules*? It's kinda what you do in a crisis. You have to break all the rules. What you do: knock down the bureaucracy, get rid of all the committee meetings, look at your policies and procedures that don't work in an emergency. All banks are required to have a disaster recovery plan, and ours is thick. If you can find the sucker, you'll be lucky after a storm, but you sure don't have time to go read the rules you were making about. What if? What if? What if? You just have to make decisions and get on with the job. And that's what Travnicek did. He didn't do anything wrong. I don't know whether he was in communication with the state health officer or not, if he could even talk to the state health officer. He was the epicenter of what happened health-wise, and he made it happen."

§

Deputy commander of the initial Incident Management Team deployed to Harrison County, Martin Senterfitt declared, "It's one thing for sure—we're flattered that they appreciated us and our help, but the difference between my people, Chip Patterson's people, who went into Mississippi and the people who were already there is simply this: we were not disaster victims. My wife and my kids were sitting in air conditioning, watching television, and my home was still safe. We were able to come into Mississippi with a certain level of detachment; we could stay above the emotion of it, whereas people like Bobby, Rupert, Dr. Travnicek, General Spraggins—this was their home. These were their friends, and yet they were still able to function at such a high level. A number of times, and it was almost surreal, we would be in the middle of a day working, and all of a sudden we would see groups of locals pull off together, whispering to each other. And actually they'd all start crying together and hugging each other. And they would stay about five minutes, crying and hugging each other, and then they all would go back to work. It finally hit me: they're hearing news. 'Hey, remember our coworker? Yeah, okay. They just pronounced her dead. They found her body.' Or 'Hey—do you remember' this man or that woman? These were people who still didn't know where all their family was. These were people whose coworkers were dead; their family members were dead; their homes were destroyed. And yet they were still right there in that county EOC working 18 hours a day and getting the job done.

"I get very frustrated, to the point of anger, righteous anger, when I hear people talking negatively about the Katrina response. The truth of the matter is—I can only speak for what I saw in Harrison County, Mississippi—but they were some of the most courageous, heroic people I've ever seen in my life. This was their home. Their homes were destroyed, and yet they stuck up, and they did the right thing. We've got to make sure that that's where the praise goes. Nine days into it, I was absolutely, purely exhausted. To know that all those people in Harrison County continued to fight the good fight for weeks, months, and years, is just an incredible story. I just can't say enough about the hard work that they gave."

§

Dr. Bob Travnicek weathered Katrina, also known on the Gulf Coast as the Bitch Storm, with Mississippi colleagues who thought they had already experienced the worst-of-the-worst in 1969 at the brunt of Category 5 Hurricane Camille. They had not. Katrina clobbered Mississippi in 2005. Already under fire from the state level and under the scrutiny of what he considered an incompetent state health officer and conflicted Board of Health, the physician had his hands full dealing with daily pressures of a six-county public health district. Beyond, he shouldered responsibility for public health and medical services preparedness and response. His was a game of alliances and distractions, suffering and survival.

"I have always been interested in why things work, and why things don't work," Travnicek said. "And because I value people, it's all about people. Just as I stood in the middle of the storm, I thought it was great fun! This is the world's highest stakes game, public health. Somebody got mad at me when I first came down here, and they picketed me in a windstorm, and the media were there. One of the picket people was a little old lady, and she wasn't very strong. While the television cameras were rolling, the sign got away from her, and it hit me on the head, knocked me down. The reporter comes down, and she takes the thing, and she said, 'Dr. Travnicek, how do you feel?' I said, 'I feel great.' And she said, 'Why would you feel great?' I said, 'I have a chance to make a difference; this is the lowest point, and we're coming back,' you know?

"I don't see things in terms of negatives. I may win, I may lose. I'm passionate about what I believe. One of the things Dr. Amy said was that he had heard I was the most passionate health officer in the state, for sure, but he thought I had carried that way too far. Told me that the first day. I thought about that and said, 'Gosh, let me think about that.' You know— what are you going to do? I don't know if you've done any big game hunting, but I've been all over the world, and I have been in every kind of difficult and scary situation you can think of, and I can tell you this: when the tiger has his paw on your neck, you do not move."

Epilogue – 2015

Brigadier General Joe Spraggins had given up his active duty slot and moved into a dual role as chief of staff, Tennessee Air National Guard, and battle commander, First Air Force, before taking the Harrison County Emergency Management Director position in August 2005. Approaching retirement and planning his next move, he had asked God for a challenge. When Harrison County offered the position, he truly believed "that's what the Lord wanted me to do. I had asked the Lord for a challenge in my next career, so now I 'qualify' challenge. He gave me the challenge of a lifetime, but He gave me the will and the ability to do it. It's like whenever I would go to speak to the media, I would stop and say a little prayer—*'Lord, just take the words you want me to say and put them into my mouth'* – and there's a lot of times—I'm not lying—words came out of my mouth that I had no idea where they came from. They were from Him, and it was the right thing."

Hurricane Katrina slammed the Mississippi Gulf Coast on what was to have been Spraggins' first day on the job, August 29, 2005. Approaching the 10-year anniversary of the catastrophic storm, Spraggins still considers "the devotion and willpower of South Mississippi and the people" to be the biggest thing Katrina delivered. "When everybody became equal— no matter how much money you had, or not, everybody chipped in. The richest people to the poorest people chipped in together and said, 'Hey! We're here to help.' To see that working in America, with that many people coming together and working together. . . We had very wealthy people working in those distribution centers, handing out stuff, and we had very wealthy people going through there to get something, too. It equalized us to a point—it didn't matter who you were or what you were or your status or anything else, we were all South Mississippians. And we had to move together. I think it proves that when something happens, the people of South Mississippi will pull together and make it happen."

In numerous presentations before completing his work in emergency management, returning to the private sector, and becoming chief operating officer of Mississippi Department of Marine Resources in 2013, Spraggins admonished all who aspire to "get ready" for the unknown: be prepared for the worst, be prepared for a long recovery, have a plan for everything, ask for help, make a decision, believe in the Incident Command System "because it works."

§

Throughout Katrina's destruction and the county's response to her impact, as the Gulf Coast advanced toward recovery and rebuilding, District Health Officer Robert Travnicek tried to ignore the people he knew as political pawns and continued to focus on medical, public health, and environmental healing. He knew that the cure had to start with individuals; he *would* escape from the under-the-guillotine stress he personally suffered through the ordeal at the hands of a disrespected director of Mississippi State Department of Health. The first glimmer of relief came on November 8, 2005, when the State Legislature's Performance Evaluation and Expenditure Review (PEER) Committee issued its "limited management review" of the department. PEER criticized the agency for mismanagement, failure to involve the board in structural reorganizations, restriction of internal and external communications, not using model resource allocation and performance measurement, and loss of public health experience and knowledge.

Subsequent critics blamed State Health Officer Amy and the then-divided board for broad-based public dissatisfaction of state health department operations, including multiple reorganizations, a bloated top-management tier at headquarters and dwindling workforce at the local level statewide, and dramatic increases in infectious diseases and the infant mortality rate.

Travnicek and his team on the Coast continued with traditional public health programs operations, and he with his key district-level staff worked tirelessly with local officials to rebuild the infrastructure. He knew water and sewers to be the most important consideration for a healthy community in the future.

With PEER's conclusion that Mississippi's once-lauded public health system no longer measured up and in light of increasing media attention, then-Senate Public Health Committee Chairman Alan Nunnelee convened three rounds of public investigative hearings in August, September, and November 2006—just a year after Katrina. Former State Health officers Alton B. Cobb and Ed Thompson, previous and then-current department employees, board members, and healthcare experts testified. Amy and the board denied all allegations, and he doggedly maintained that "public health is about people; we are in the business of customer service. The agency has set forth on an ambitious new program of setting goals, measuring results, and improving our performance in the pursuit of public health excellence."

The chaotic year of unheralded public scrutiny and Senate Public Health Committee's "no confidence vote" for Amy led legislators to craft Senate Bill 2764 in their 2007 Session. With Nunnelee's leadership and help from House Public Health Committee Chairman Steve Holland and their colleagues, Mississippi got a totally restructured State Board of Health. During the 2007 Session, legislators eliminated the previous board, deleted the state health officer position, and prescribed new direction for a revamped board and Department of Health and also solidified qualifications for the executive director and duties of the board and agency. The "sunset" law effectively unseated Amy.

Cobb, the physician who led the public health system from 1973 through 1992 commented: "The basic problem was the board. Now that we're over that hurdle; we've got a fresh start. I feel very positive. For the executive director of the agency to have been removed from office by the legislature is a first—definitely the first time in state history that both a board and agency head have been removed."

§

Travnicek did not talk about it publicly until years later. Everything he was and knew from studying and practicing both medicine and public health had helped him not only survive but also lead public health emergency preparedness and response to Hurricane Katrina. In 2012, Mississippi State Medical Association affirmed his "legendary service and dedication to public health" by presenting to him its Physician Award for Community Service. Including a cash award to the civic organization of his choice, the recognition honors both active practice of medicine and community service above and beyond the call of duty. Accepting the plaque, Travnicek talked for only the second or third time in seven years about the storm that immediately and forever changed lives and the Mississippi Gulf Coast.

The fellow physician who proposed the award called the recognition overdue but apt for "a long and illustrious medical career which has spanned nearly five decades. He is a man of highest character and deserves recognition of this prestigious award." Responsibility for a broad swath of public health concerns and a securely-developed position of authority throughout the district empowered the public health doctor to immediately respond to Hurricane Katrina.

In crediting others, Travnicek personified the strengths of his public health career: team-building and cheerleading. Accepting the award, he said, "This gives me a chance for the first time to publicly recognize at least three of the hundreds with whom I was teamed." He lauded George Schloegel, then president of Hancock Bank who became Mayor of the City of Gulfport; Gary Marchand, CEO of Gulfport Memorial Hospital; and "finally and particularly, my long-suffering wife, who, as a staff nurse, was in a lockdown situation at Garden Park Hospital during the storm and for two days after."

Surviving Katrina as a staff nurse in a partially destroyed and flooded hospital, Lora Travnicek days later went home to a "mostly destroyed home, with the roof effectively ripped off. With a friend, she cut and tacked up what was left of our pool enclosure to form a screen door" and then "fed the generator and slept with a shotgun between her legs for three weeks until we finally got restoration of our electricity."

Schloegel, Travnicek praised, put cash immediately back into the hands of storm victims—some $30 million worth of cash from the Federal Reserve in Atlanta.

"His 15-story flagship bank was functionally destroyed for 18 months. He ordered $3 million in small denomination bills and handed them out at storefront offices and even on the street to anyone who would sign an IOU. All of this was against a backdrop of utter destruction: no communication, no water, no sewer, no electricity. He immediately allowed people, most of whom had lost everything, some sense of hope in a hopeless situation."

A few blocks away from the bank's location, Gulfport Memorial sustained substantial structural damage but sheltered and fed about 600 families—staff plus countless members of the community who were rendered homeless until public shelters could be activated or those trapped by the storm could leave the area. The hospital operated on only emergency power for almost a week, and health care teams performed cesarean sections by flashlight. With no help from federal responders, Marchand managed to find and burn diesel fuel to keep essential equipment powered and the hospital open.

That was only the first of Travnicek's many awards and honors; his most-appreciated would follow in January 2014 when the City of Gulfport renamed a street Dr. Robert Travnicek Boulevard. The new name for 45th Avenue between West Railroad Street and Memorial Drive in Gulfport celebrates two dozen years of service the public health physician devoted to residents of the city and county through June 2013. Unveiling took place at the Harrison County Health Department, opened in 2003 and marking the end of a 13-year building effort. Travnicek noted the value of working with local elected officials to build the facility to withstand Category 5 hurricane conditions. He recalled that the county building was ready to re-open the day after Hurricane Katrina smashed into the Mississippi Gulf Coast. No other building in the vicinity could claim that. And no other known Mississippi health officer can claim a public street's name.

When Travnicek noted the street's length to be quite short, his friend and fellow Katrina champion, Dr. Ray Basri of New York, heartily congratulated him: "Not even the Champs-Élysées would be long enough to truly reflect your brilliance and spirit."

Harrison County Youth Court Judge Margaret Alfonso prompted the renaming, which the City of Gulfport unanimously approved. In their proclamation, Travnicek gets credit for having "championed needed change and created plans and policies to provide expanded services for the residents." Proclamation writers credited Travnicek for having "worked tirelessly in the aftermath of Hurricane Katrina to ensure that the public health disaster created by the wind and waves of the hurricane was limited to the greatest extent possible, keeping the health of the citizens at the forefront of his efforts." They cited his "knowledge, experience, compassion, empathy, and determination" to expand "the role and benefit of his office in a manner which has enhanced the prevention and treatment of threats to the health of Harrison County's residents and visitors."

Under his own steam and at his own time, Travnicek retired from public service in June 2013.

§

Martin Senterfitt, then a 24-year veteran, became director and fire chief, City of Jacksonville, Florida, in December 2011.

"An event like that creates far more emotional baggage than you even realize. For the first two or three years after that event, when I started talking about it, I'd tear up and choke up. Or I'd get angry at some of the things. It's been 10 years now; I'm beyond the anger. I can now usually control the emotions. I'll tell you one of the things that just floors me. The Northeast Florida team was given a lot of credit for coming in and helping Harrison County organize—what we really did was organize them and give them some shelter to get back on their feet. I remember the team there in Mississippi looked right at us when we went to leave and said, 'Hey. One thing we know in Mississippi is we pay our debts when somebody's helped us; so if you ever need us, we will be there.' And, of course, several of us were from Jacksonville. . . I can tell you, since 2005, we've been threatened by four different tropical storms. Every single time a hurricane starts threatening Jacksonville, Harrison County, Mississippi—Rupert Lacy—will call us one or two days before the storm and say, 'We're ready to respond; if you need us, say the word, and we're on the way.' That's what makes this country great, and that's what makes this system work. The people who are not being impacted have to stand up and help the people that are being impacted. When you do that, good things happen to all of us."

§

Robert Latham, who led Mississippi through Katrina, retired as executive director of the Mississippi Emergency Management Agency in 2006, worked as an independent consultant in emergency management and homeland security for nearly six years, and accepted Governor Phil Bryant's appointment for his second stint as executive director of MEMA in 2012. He sees weather events in the Deep South in 2005 as "two different stories. One is Hurricane Katrina, which made landfall in Mississippi; the other one is the levee failure in New Orleans—yes, it was caused by Katrina, but they did not get a hurricane. They were on the west side of the storm. Katrina made landfall in Hancock County. That part of the story too often gets lost."

He keeps a copy of *Hurricane Katrina: A Nation Still Unprepared* close for ready reference. "I don't think anybody would—if they were under oath, and at the Senate hearing I did have to swear—I don't think anybody would intentionally lie. I just think that the perspective depends on where the person was at the time and what they saw and perceived. I can tell you based on what I've heard and seen, but you can talk to somebody that was in a different role or somewhere else, and you'll get a totally different story. There was so much going on. There wasn't any aspect of our daily life that wasn't impacted."

Among originators of Mississippi's unified command structure to manage the Katrina catastrophe, Latham says the operation evolved. "It's really not a command, it's a UCG, which is a unified coordination group. You have public health at the table, Department of Public Safety at the table, health and human services—all the players at the table, and everybody is equal. There was no, 'I'm in charge,' and that's why it worked. It evolved from the two-I'ed NIIMS to the one-I'ed NIMS, then to unified command and, ultimately, to Unified Command Group. It's all connected. Focus on the UCG as built on the incident command structure. What's important is that everybody who had a particular role had a seat at the table; so it's not like one or two people sitting in this office deciding and then going out to tell everybody, 'We're going to do this.' The group would develop an incident action plan, and it was a joint plan between the state and the federal government. For the next 24 hours, here are our priorities, here are the resources, communications plan—if you want to focus on the success, you can go back to the Executive Order making us the sixth state to do that and then to UCG, the first time it had been done successfully in a major disaster in this country. It was done in Mississippi."

Latham is among many trained to look back so that he can see ahead. From what Katrina did to and left in Mississippi, he predicts another "disaster that probably has started but might not be realized for several more years, and that's the mental health disaster. At the peak, we had 48,000 families in travel trailers. So, for a lot of children, a significant part of their formative years, of their developmental years, was spent in a very small area trailer. There is funding in every disaster for crisis counseling, but it doesn't typically deal with treatment; it just identifies it. I think there is probably right under the surface a mental health crisis—not only the adults but the children who will remember their childhood in a travel trailer because their home was destroyed. I haven't seen anything, but I would suspect that would be another crisis, lack of mental health counseling immediately after and among the first responders and people who live there and are still dealing with it today. They are the obvious victims, for me. After I got back and ran on adrenalin for a certain period of time and then—you just run into a brick wall. That happened to me a couple months after I'd gotten back up here. I was working up here every day; there wasn't really a time to start the day and a time to end the day. It was all 24 hours. We were all working shifts, and I remember getting into the car to drive to work one day, and I just broke down. For no apparent reason. I think it just caught up with me.

"And the thing that amazes me is that—I don't know how those people who live on the Coast get up and go every day. Think about what they've been through. How do you do that? How? It would be just real easy to say, 'I can't do this anymore,' and move north. But those people are resilient."

§

Hancock Bank President George Schloegel had intended to retire in December 2005, but Hurricane Katrina delayed that plan for three years. "Then I did a crazy thing; I ran for mayor and served 2009 through 2013."

In 2015, he continues to count the positives Mississippi claimed then and into the future. "We are very fortunate on the Mississippi Gulf Coast because we're the headquarters for the Atlantic fleet Seabees, one of only two in the country, Port Hueneme, California, for the Pacific fleet and Gulfport, Mississippi, for the Atlantic. And we're the technology training center for the Air Force at Keesler. In addition to that, we have a strong Mississippi

National Guard, both Army as well as Air Force, and those forces were commandeered quickly to come down and help where we needed the help. The president of the United States was here personally a number of times to give whatever assistance he and his agencies could give. That means an awful lot. And I'm sure they broke some rules. I heard one of the directors from FEMA a year after the storm, and he said, 'Boy, did we learn a lot.' And I went onto what they call the National Advisory Board for FEMA right after the storm; they were telling us that when the storm hit, they had something like 1,700 FEMA employees in the whole country and within 60 days of the storm, that swelled to something like 14,000 people. How do you build an organization that big that quickly without having a lot of problems? But that's what they did. That's the first time FEMA had *really* been put to the test on a large scale. How do you get that many additional employees and mobilize them, and send them where they need to go to get something done and to crack through all the federal bureaucracy to make things happen?

"I asked one of the FEMA people when I was on that board. 'Tell me something: we spent billions of dollars. How much of the money we spent, do you think, was wasted? How much of the money that we spent do you think was inappropriately spent?' And he said, 'About half.' I asked how he feels about that. He said, 'Pretty good.' He said, 'When you have a calamity that large, you've got to forget the mistakes that you make because that's the cost of getting the other things that were right done. You just have to accept the fact that inefficiency is built in, but don't let that stop you from doing the things that you did right.' That's a pretty good statement. I've got a slide I put up when I do my banking school: Done Beats Perfect. There is no perfect."

§

Rupert Lacy stepped into new, increasingly-familiar boots as Harrison County Emergency Management Agency deputy director in August 2006, became interim director nine months later, and took the full-time director appointment in December 2007.

In many ways still responding to Katrina, he assures with certainty, "The Coast is much better prepared for another monster. Now we all take possible tropical storms and hurricane warnings seriously. Our residents prepare;

they buy their supplies—we don't see the mad rush we saw before Katrina."

And the people who do need a safe shelter away from their home can count on hurricane built-to-purpose storm shelters designed to withstand powerful winds and located outside of flood-prone areas. The facilities are equipped to meet general population or special medical needs. "The resident can stay within his or her own community, ride the storm out, and then get back home, check on properties, and do what needs to be done."

The federal government's investment of more than five billion dollars into Mississippi allowed the Gulf Coast to "get a makeover after Katrina, and that makes a world of difference. We can recover from future storms more quickly and with less hassle because we have built the infrastructure—the water, sewer, and power grids—to higher standards. Our new residential and commercial construction meets standards we learned from Katrina. For the future, we have tried to prepare for the worst and pray for the best."

Resources

Brinkley, Douglas. *The Great Deluge: Hurricane Katrina, New Orleans, and the Mississippi Gulf Coast.* New York: William Morrow, An Imprint of Harper Collins Publishers, 2006. Print.

Harrison County Emergency Management Agency. (2005) [Press Releases]

Hearn, Philip D. *Hurricane Camille: Monster Storm of the Gulf Coast.* Jackson: University Press of Mississippi, 2004. Print.

Heroes of Katrina [Motion picture on DVD]. (2006). USA: Memorial Hospital of Gulfport.

Smith, James Patterson. *Hurricane Katrina: The Mississippi Story.* Jackson: University Press of Mississippi, 2012. Print.

Special Report of the Committee on Homeland Security and Governmental Affairs. *Hurricane Katrina: A Nation Still Unprepared.* Washington: US Government Printing Office, 2006. Print.

(2005). *The Clarion-Ledger.*

http://www.c-span.org/video/?190224-1/hurricane-katrina-preparedness

http://www.cdc.gov

http://www.cdcfoundation.org

http://www.defense.gov/news/newsarticle.aspx?id=61033

http://www.dhs.gov

http://www.fema.gov

https://www.fema.gov/news-release/2011/08/29/six-years-after-hurricane-katrina-mississippi-continues-recover-and-rebuild

http://www.fema.gov/pdf/about/pub1.pdf

http://www.fema.gov/txt/nims/nims_ics_position_paper.txt

http://www.floridadisaster.org

http://georgewbush-whitehouse.archives.gov/reports/katrina-lessons-learned

https://www.google.com/maps

http://www.gulfcoast.org/visitors/attractions/beaches-and-harbor-activities

http://www.gulfportmemorial.com

https://www.hancockbank.com

http://www.history.com/topics/hurricane-katrina

http://www.hmc.org

http://www.hurricane.com

http://www.katrina.com

http://www.katrina.house.gov

http://www.katrina.house.gov/hearings/12_07_05/witness_list_120705.htm

http://www.livescience.com/22522-hurricane-katrina-facts.html

http://msdh.state.ms.us

http://msdh.ms.gov/msdhsite/_static/resources/1676.pdf

http://www.msema.org

http://www.msstateguard.org

http://www.nasa.gov/centers/stennis/home/#.VRwQdY63tD0

http://www.ncdc.noaa.gov/oa/reports/tech-report-200501z.pdf

http://www.nhc.noaa.gov

http://www.nhc.noaa.gov/archive/2005/pub/al122005.public.021.shtml

http://www.noaa.gov

http://www.noaanews.noaa.gov/stories2012/20120530_eyewall.html

http://www.nytimes.com/2005/09/01/business/01oil.html?pagewanted=all&_r=0

http://www.seabeesmuseum.com/history.html

http://www.weather.com

https://www.youtube.com/watch?v=eHcyC5sc_so

https://www.youtube.com/watch?v=FfoR8EuSQLk

https://www.youtube.com/watch?v=0tqmPCs07ko

Index

Made in the USA
Middletown, DE
23 July 2016